SHOE and LEATHER

ENCYCLOPEDIA

A Book of Practical and Expert Testimony by Successful Merchants

Each Article a Chapter
Each Chapter a Single
and Separate Subject

A History of Shoemaking

Shoemaking, at its simplest, is the process of making footwear. Whilst the art has now been largely superseded by mass-volume industrial production, for most of history, making shoes was an individual, artisanal affair. 'Shoemakers' or 'cordwainers' (cobblers being those who repair shoes) produce a range of footwear items, including shoes, boots, sandals, clogs and moccasins – from a vast array of materials.

When people started wearing shoes, there were only three main types: open sandals, covered sandals and clog-like footwear. The most basic foot protection, used since ancient times in the Mediterranean area, was the sandal, which consisted of a protective sole, attached to the foot with leather thongs. Similar footwear worn in the Far East was made from plaited grass or palm fronds. In climates that required a full foot covering, a single piece of untanned hide was laced with a thong, providing full protection for the foot, thus forming a complete covering. These were the main two types of footwear, produced all over the globe. The production of wooden shoes was mainly limited to medieval Europe however – made from a single piece of wood, roughly shaped to fit the foot.

A variant of this early European shoe was the clog, which were wooden soles to which a leather upper was attached. The sole and heel were generally made from one piece of maple or ash two inches thick, and a little longer and broader than the desired size of shoe. The outer side of the sole and heel was fashioned with a long chisel-edged implement, called the clogger's knife or stock; while a second implement, called the groover, made a groove around the side of the sole. With the use of a 'hollower', the inner sole's contours were adapted to the shape of the foot. In even colder climates, such designs were adapted with furs wrapped around the feet, and then sandals wrapped over them. The Romans used such footwear to great effect whilst fighting in Northern Europe, and the native Indians developed similar variants with their ubiquitous moccasin.

By the 1600s, leather shoes came in two main types. 'Turn shoes' consisted of one thin flexible sole, which was sewed to the upper while outside in and turned over when completed. This type was used for making slippers and similar shoes. The second type united the upper with an insole, which was subsequently attached to an out-sole with a raised heel. This was the main variety, and was used for most footwear, including standard shoes and riding boots.

Shoemaking became more commercialized in the mid-eighteenth century, as it expanded as a cottage industry. Large warehouses began to stock footwear made by many small manufacturers from the area. Until the nineteenth century, shoemaking was largely a traditional handicraft, but by the century's end, the process had been almost completely mechanized, with production occurring in large factories. Despite the obvious economic gains of mass-production, the factory system produced shoes without the individual differentiation that the traditional shoemaker was able to provide.

The first steps towards mechanisation were taken during the Napoleonic Wars by the English engineer, Marc Brunel. He developed machinery for the mass-production of boots for the soldiers of the British Army. In 1812 he devised a scheme for making nailed-boot-making machinery that automatically fastened soles to uppers by means of metallic pins or nails. With the support of the Duke of York, the shoes were manufactured, and, due to their strength, cheapness, and durability, were introduced for the use of the army. In the same year, the use of screws and staples was patented by Richard Woodman. However, when the war ended in 1815, manual labour became much cheaper again, and the demand for military equipment subsided. As a consequence, Brunel's system was no longer profitable and it soon ceased business.

Similar exigencies at the time of the Crimean War stimulated a renewed interest in methods of mechanization and mass-production, which proved longer lasting. A shoemaker in Leicester, Tomas Crick, patented the design for a riveting machine in 1853. He also introduced the use of steam-powered rolling-machines for hardening leather and cutting-machines, in the mid-1850s. Another important factor in shoemaking's mechanization, was the introduction of the sewing machine in 1846 – a development which revolutionised so many aspects of clothes, footwear and domestic production.

By the late 1850s, the industry was beginning to shift towards the modern factory, mainly in the US and areas of England. A shoe stitching machine was invented by the American Lyman Blake in 1856 and perfected by 1864.

Entering in to partnership with Gordon McKay, his device became known as the McKay stitching machine and was quickly adopted by manufacturers throughout New England. As bottlenecks opened up in the production line due to these innovations, more and more of the manufacturing stages, such as pegging and finishing, became automated. By the 1890s, the process of mechanisation was largely complete.

Traditional shoemakers still exist today, especially in poorer parts of the world, and do continue to create custom shoes. In more economically developed countries however, it is a dying craft. Despite this, the shoemaking profession makes a number of appearances in popular culture, such as in stories about shoemaker's elves (written by the Brothers Grimm in 1806), and the old proverb that 'the shoemaker's children go barefoot.' Chefs and cooks sometimes use the term 'shoemaker' as an insult to others who have prepared sub-standard food, possibly by overcooking, implying that the chef in question has made his or her food as tough as shoe leather or hard leather shoe soles. Similarly, reflecting the trade's humble beginnings, to 'cobble' can mean not only to make or mend shoes, but 'to put together clumsily; or, to bungle.'

As is evident from this short introduction, 'shoemaking' has a long and varied history, starting from a simple means of providing basic respite from the elements, to a fully mechanised and modern, global trade. It is able to provide a fascinating insight not only into fashion, but society, culture and climate more generally. We hope the reader enjoys this book.

INTRODUCTION

THE retail shoe merchant who reads a copy of this booklet will realize how much solid information every chapter contains.

The keen rivalry seen on every hand in the merchandise world to-day clearly demonstrates the fact that the merchant who will succeed must take advantage of every help that comes his way.

Store methods which were successful a score of years ago find no place in the store of today, and the merchant who enjoys the trade in his community must keep in constant touch with the wishes and suggestions of his patrons.

Some minute detail which has escaped his notice may be the cause of losing many dollars' worth of business. No matter how systematic the merchant may be, nor how attractive a line he handles, nor how efficient his help, he must keep constantly on the lookout to follow the wishes of his customers and anticipate them, if such a thing is possible.

Every article in this book was written by men who have been actively engaged in the retail shoe business. Give the book the consideration which it deserves, and get the benefit of their experiences. They are thoughtful, discriminating men and good merchants.

This book is presented to the new subscribers of the GAZETTE with the best wishes of the donors. You are cordially invited to write us at any time concerning your methods of buying or selling or advertising or stock-keeping, any of the problems that come up in your business. Your letters will always receive our careful consideration and an early reply.

CONTENTS

List of Named Shoes.

First Complete Collection With Address of the Manufacturer.

AARON F. SMITH, A. F. Smith Shoe Co., Lynn, Mass.
ABBOT, Lewis A. Crossett, Inc., North Abington, Mass.
ACME CUSHION, Utz & Dunn Co., Rochester, N. Y.
ADIRONDACK, Endicott, Johnson & Co., Endicott, N. Y.
ADMIRABLE, A. Priesmeyer Shoe Co., Jefferson City, Mo.
ADONIA, Nathan D. Dodge Shoe Co., Newburyport, Mass.
ADVOCATE, Nolan-Earl Shoe Co., San Francisco, California.
AEROPLANE, The Richard Gregory Shoe Co., Lynn, Mass.
AETNA WELT, Green-Wheeler Shoe Co., Ft. Dodge, Iowa.
AGATE, Friedman-Shelby Shoe Co., St. Louis, Mo.
"AGNES SCOTT", J. K. Orr Shoe Co., Atlanta, Ga.
ALAMO, Peters Shoe Co., St. Louis, Mo.
ALCIA, Daniel Green Felt Shoe Co., Dolgeville, N. Y.
ALICE, Brown Shoe Co., St. Louis, Mo.
ALICE, Nathan D. Dodge Shoe Co., Newbury, Mass.
ALL AMERICA, Rice & Hutchins, Boston, Mass.
ALL AMERICAN SPECIAL, Rice & Hutchins, Boston, Mass.
ALLEN, Friedman-Shelby Shoe Co., St. Louis, Mo.
ALLEN D., Jos. Rosenheim Shoe Co., Savannah, Ga.
ALLERTON, The Washington Shoe Mfg. Co., Seattle, Wash.
ALLIGATOR, Brown Shoe Co., St. Louis, Mo.
ALL-FOR-WEAR, Peters Shoe Co., St. Louis, Mo.
"ALLWIN", Matchless Shoe Co., Boston, Mass.
ALPHA SHOE, C. S. Marshall Co., Brockton, Mass.
ALVIN, Friedman-Shelby Shoe Co., St. Louis, Mo.
ALWAYS EASY, Peters Shoe Co., St. Louis, Mo.
ALWAYS EASY, Golden State Shoe Co., Los Angeles, Cal.
AMAZEENA HIGH QUALITY, Hayward Bros. Shoe Co., Omaha, Neb.
AMBOY, A. Priesmeyer Shoe Co., Jefferson City, Mo.
AMERICAN BEAUTY, A. H. Berry Shoe Co., Portland, Me.
AMERICAN BEAUTY, Watson Plummer Shoe Co., Chicago, Ill.
"AMERICAN BOY," Menzies Shoe Co., Detroit, Mich.
AMERICAN GENTLEMAN, Hamilton-Brown Shoe Co., St. Louis, Mo.
AMERICAN GIRL, Wolf Bros. & Co., Cincinnati, Ohio.
AMERICAN LADY, Hamilton-Brown Shoe Co., St. Louis, Mo.

AMERICAN QUEEN, Arnold-Henegar-Doyle Co., Knoxville, Tenn.
AMIGA, Friedman-Shelby Shoe Co., St. Louis, Mo.
ANATOMICAL, Nesmith Shoe Co., Brockton, Mass.
ANCHOR BRAND, Guthmann, Carpenter & Telling, Chicago, Ill.
ALABAMA, Friedman-Shelby Shoe Co., St. Louis, Mo.
ALFALFA, Friedman-Shelby Shoe Co., St. Louis, Mo.
ANGLO, Endicott, Johnson & Co., Endicott, N. Y.
ANNETTE, Daniel Green Felt Shoe Co., Dolgeville, N. Y.
ANNVILLE, A. S. Kreider Shoe Co., Annville, Pa.
ANNVILLE, The Kreider, Schwarz & Sallenbach Shoe Co., St. Louis, Mo.
ANTI-TRUST, Endicott, Johnson & Co., Endicott, N. Y.
ANTOINETTE, Daniel Green Felt Shoe Co., Dolgeville, N. Y.
APEX, The Spex Shoe Factory, New Orleans, La.
APOLLO, Dittmann Shoe Co., St. Louis, Mo.
APPLE BLOSSOM, Crowder-Cooper Shoe Co., Indianapolis, Ind.
AQUATIGHT, Rice & Hutchins, Boston, Mass.
ARCH CITY, The Henry C. Werner Co., Columbus, Ohio.
ARCH DOCTOR, Heywood Boot & Shoe Co., Worcester, Mass.
ARCH REST SHOE, Utz & Dunn Co., Rochester, N. Y.
ARCH SUPPORT, O'Sullivan Bros. Co., Lowell, Mass.
ARGOOD, Rice & Hutchins, Boston, Mass.
ARISTOCRAT, Brown Shoe Co., St. Louis, Mo.
ARIZONA, Dittmann Shoe Co., St. Louis, Mo.
ARIZONA, Hamilton-Brown Shoe Co., St. Louis, Mo.
ARM CHAIR, Brown Shoe Co., St. Louis, Mo.
ARMADA, Rice & Hutchins, Boston, Mass.
ARMORED CRUISER, Excelsior Shoe Co., Portsmouth, Ohio.
ARMSTRONG, THE, Stillman-Armstrong Co., Vanceboro, Me.
ARNOLD, Arnold Henegar Doyle Co., Knoxville, Tenn.
ARROW, Friedman-Shelby Shoe Co., St. Louis, Mo.
ARROW BRAND, Hayward Bros. Shoe Co., Omaha, Neb.
ARTCRAFT, The Tappan Shoe Mfg. Co., Coldwater, Mich.
AS-BORN, Chipman, Harwood & Co., Boston, Mass.
ASBESTOS, C. W. Johnson, Natick, Mass.
ATALUS BRAND, Milford Shoe Co., Milford, Mass.
ATHLETE, Endicott, Johnson & Co., Endicott, N. Y.
ATLANTIC, Friedman-Shelby Shoe Co., St. Louis, Mo.
ATLANTIC SPECIAL, Friedman-Shelby Shoe Co., St. Louis, Mo.
ATTRACTION, Peters Shoe Co., St. Louis, Mo.
AURORA, Peters Shoe Co., St. Louis, Mo.
AUTOGRAPH, Craddock Terry Co., Lynchburg, Va.
AUTOCRAT, Ellet-Kendall Shoe Co., Kansas City, Mo.

AVENUE, Brown Shoe Co., St. Louis, Mo.
AVISCOTT, Brown Shoe Co., St. Louis, Mo.
B. & L. SHOE, Bode-Larson Shoe Company, Keokuk, Iowa.
BABBETTE, Daniel Green Felt Shoe Co., Dolgeville, N. Y.
"BABEKIN", J. J. Goodwin, Rochester, N. Y.
BABY, Friedman-Shelby Shoe Co., St. Louis, Mo.
BABY KING, Mrs. A. R. King. Corp., Lynn, Mass.
BABY MINE, Peters Shoe Co., St. Louis, Mo.
BABY ROBIN, Northern Shoe Co., Duluth, Minn.
BACHELOR, Brown Shoe Co., St. Louis, Mo.
BADGER BOY, Excelsior Shoe & Slipper Co., Cedarburg, Wis
"BAKER", Fox Baker Co., Rochester, N. Y.
BAKER & BOWMAN, Syracuse Shoe Mfg. Co., Syracuse, N. Y
BANISTER, J. A. Banister Co., Newark, N. J.
BANNER, Huiskamp Bros. Co., Keokuk, Iowa.
BAREFOOT, Peters Shoe Co., St. Louis, Mo.
BAR HARBOR, Friedman-Shelby Shoe Co., St. Louis, Mo.
"BARKER", Barker, Brown & Co., Huntington, Ind.
BARKER'S BEVELED TAP, Barker, Brown & Co., Hunting-
    ton, Ind.
BARNEY OLDFIELD, Roberts, Johnson & Rand Shoe Co.,
    St. Louis, Mo.
BARNYARD, Huiskamp Bros. Co., Keokuk, Iowa.
BARTON SHOE, Barton Bros., Kansas City, Mo.
"BATES," A. J. Bates Co., Webster, Mass.
BEACON SHOES, F. M. Hoyt Shoe Co., Manchester, N. H.
BEATRICE, Dittmann Shoe Co., St. Louis, Mo.
BEATS ALL, Brown Shoe Co., St. Louis, Mo.
BEATS ALL, The Cady-Ivison Shoe Co., Cleveland, Ohio.
BEAUMONT, Friedman-Shelby Shoe Co., St. Louis, Mo.
BEAUTY, Peters Shoe Co., St. Louis, Mo.
BEAUTY BOOTS, Blum Shoe Mfg. Co., Danville, N. Y.
BEAVER, Friedman-Shelby Shoe Co., St. Louis, Mo.
BED ROCK, Foot, Schulze & Co., St. Paul, Minn.
BELLE OF BOSTON, Crowder-Cooper Shoe Co., Indianapolis,
    Ind.
BENCH MADE, Geddes-Brown Shoe Co., 215 S. Meridian St.,
    Indianapolis, Ind.
BENCH-MADE, Geo. G. Snow Co., Brockton, Mass.
BENCH WORK WHITE HOUSE, Brown Shoe Co., St. Louis,
    Mo.
BEND EASY, Xenia Shoe Mfg. Co., Xenia, Ohio.
BEND WELL, C. P. Ford & Co., Rochester, N. Y.
BENSON, Hamilton-Brown Shoe Co., St. Louis, Mo.
BERNALDA, Thomson-Crooker Shoe Co., Boston, Mass.
BEST, Slater & Morrill, Inc., South Braintree, Mass.
BEST BOY, The Kreider, Schwarz & Sallenbach Shoe Co., St.
    Louis, Mo.

BEST BOY, A. S. Kreider Shoe Co., Annville, Pa.
BEST YET, Friedman-Shelby Shoe Co., St. Louis, Mo.
BEST YET, Peters Shoe Co., St. Louis, Mo.
BESTYET, Daniel Green Felt Shoe Co., Dolgeville, N. Y.
BIG CHIEF, Excelsior Shoe & Slipper Co., Cedarburg, Wis.
BIG HIT, Whittinghill-Harlow Shoe Co., St. Joseph, Mo.
BIKE COMFORT, Rice & Hutchins, Boston, Mass.
BILLY-BUSTER, The Washington Shoe Mfg. Co., Seattle, Wash.
BILTMORE. Brown Shoe Co., St. Louis, Mo.
BILTRITE, Smith-Briscoe Shoe Co., Inc., Lynchburg, Va.
BIPLANE, Endicott, Johnson & Co., Endicott, N. Y.
BISMARK, Hamburger Bros. Shoe Co., St. Louis, Mo.
BLACK CAT, Wood & Johnson Co., Rochester, N. Y.
BLACK DIAMOND, A. M. Miller & Co., Orwigsburg, Pa.
BLACK DIAMOND, Peters Shoe Co., St. Louis, Mo.
BLACK ELK, Friedman-Shelby Shoe Co., St. Louis, Mo.
BLACK HAWK, Hamilton-Brown Shoe Co., St. Louis, Mo.
BLACK RIVER, Dittmann Shoe Co., St. Louis, Mo.
BLAINE, Hamburger Bros. Shoe Co., St. Louis, Mo.
BLUE BLOOD, Golden State Shoe Co., 261 S. Los Angeles St., Los Angeles, California.
BLUE DIAMOND LINE, H. B. Turner & Co., Chillicothe, O.
BLUE & GOLD, Nolan-Earl Shoe Co., San Francisco, Cal.
BLUE MOON, James Clark Leather Co., St. Louis, Mo.
BLUE STEEL, Hamilton-Brown Shoe Co., St. Louis, Mo.
BOARDED CALF, Dittmann Shoe Co., St. Louis, Mo.
BOHEMIA, A. Priesmeyer Shoe Co., Jefferson City, Mo.
"BOND ST.", Ellis F. Copeland & Son, Brockton, Mass.
BONDORN, H. A. Noyes, Haverhill, Mass.
BONITA, A. Priesmeyer Shoe Co., Jefferson City, Mo.
"BONITA", Adler, Martin & Katz, Rochester, N. Y.
BONNE, A. Priesmeyer Shoe Co., Jefferson City, Mo.
BOOTEE, Hamilton-Brown Shoe Co., St. Louis, Mo.
BOOTEE, Friedman-Shelby Shoe Co., St. Louis, Mo.
BOSTONIANS, The Commonwealth Shoe & Leather Co., Boston, Mass.
BOULEVARD, Brown Shoe Co., St. Louis, Mo.
BOX CALF, Herold Bertsck Shoe Co., Grand Rapids, Mich.
BOX CALF, Friedman-Shelby Shoe Co., St. Louis, Mo.
BOYDEN'S SHOES, Boyden Shoe Mfg. Co., Newark, N. J.
BOY PROOF, Hamilton-Brown Shoe Co., St. Louis, Mo.
BOY SCOUTS, Excelsior Shoe Co., Portsmouth, Ohio.
BRADLEY, Friedman-Shelby Shoe Co., St. Louis, Mo.
BRAVO, Friedman-Shelby Shoe Co., St. Louis, Mo.
BRISTOL, Peters Shoe Co., St. Louis, Mo.
BROADWAY, Brown Shoe Co., St. Louis, Mo.
BRONCO, Marion Shoe Co., Marion, Ind.

"BROWNIE", Lynchburg Shoe Co., Inc., Lynchburg, Va.
BROWNWOOD, Friedman-Shelby Shoe Co., St. Louis, Mo.
BRUT TAMER, Brown Shoe Co., St. Louis, Mo.
BUCKEYE, Hamilton-Brown Shoe Co., St. Louis, Mo.
BUCKSKIN, Friedman-Shelby Shoe Co., St. Louis, Mo.
BUD'S CHOICE, The Apex Shoe Factory, New Orleans, La.
BUFFALO CALF, Bentley & Olmsted Co., Des Moines, Iowa.
BUFFALO CALF, Hayward Bros. Shoe Co., Omaha, Neb.
BULL EYE, Brown Shoe Co., St. Louis, Mo.
BULLY, Endicott, Johnson & Co., Endicott, N. Y.
BUNION, Heywood Boot & Shoe Co., Worcester, Mass.
BUNKER HILL, P. Cogan & Son, Stoneham, Mass.
BUNNY, Brown Shoe Co., St. Louis, Mo.
BURT, EDWIN C., Edwin C. Burt & Co., Brooklyn, N. Y.
BUSTER BROWN, BLUE RIBBON, Brown Shoe Co., St.
    Louis, Mo.
BUSY BEE, Sharood Shoe Corporation St. Paul, Minn.
BUTTE, Friedman-Shelby Shoe Co., St. Louis, Mo.
BUTTERFLY, Brown Shoe Co., St. Louis, Mo.
BUTTERFLY, Peters Shoe Co., St. Louis, Mo.
BUTTERFLY, I. P. Farnum, Chicago, Ill.
CREOLE, H. Lindenberg, New Orleans, La.
CRESCENT, Ellet-Kendall Shoe Co., Kansas City, Mo.
CRESCENT, A. Priesmeyer Shoe Co., Jefferson City, Mo.
CRESCENT, Hayward Bros. Shoe Co., Omaha, Neb.
CRICKET, A. Priesmeyer Shoe Co., Jefferson City, Mo.
CRITERION, Barton Bros., Kansas City, Mo.
CRITERION, Guthmann, Carpenter & Telling, Chicago, Ill.
CROSS, John H. Cross Co., Boston, Mass.
CROSSETT, Lewis A. Crossett, Inc., North Abington, Mass.
CROWN, Friedman-Shelby Shoe Co., St. Louis, Mo.
CROWN PRINCE, Bode-Larson Shoe Co., Keokuk, Iowa.
CRUSIERS "WASHINGTON", The Washington Shoe Mfg.
    Co., Seattle, Wash.
CACTUS, Hamilton-Brown Shoe Co., St. Louis, Mo.
CADET, Hayward Bros. Shoe Co., Omaha, Neb.
CADY'S CLEVELAND, The Cady-Ivison Shoe Co., Cleveland,
    Ohio.
CADY'S COLLEGE, The Cady-Ivison Shoe Co., Cleveland, O.
CADY'S EUCLID, The Cady-Ivison Shoe Co., Cleveland, O.
CADY'S MODEL, The Cady-Ivison Shoe Co., Cleveland, O.
CAESAR, Hamburger Bros. Shoe Co., St. Louis, Mo.
CALENDAR, Huiskamp Bros. Co., Keokuk, Iowa.
CALIFORNIA PRIDE, Golden State Shoe Co., Los Angeles,
    California.
CAMEL, Dittmann Shoe Co., St. Louis, Mo.

CAMEO, Rice & Hutchins, Boston, Mass.
CAPITAL, The Henry C. Werner Co., Columbus, Ohio.
CAPITOL, Golden State Shoe Co., Los Angeles, California.
CAPTAIN, Marion Shoe Co., Marion, Ind.
CARNATION, Geddes-Brown Shoe Co., Indianapolis, Ind.
CARNATION, A. Priesmeyer Shoe Co., Jefferson City, Mo.
CARRIE, Friedman-Shelby Shoe Co., St. Louis, Mo.
CASCADE, Friedman-Shelby Shoe Co., St. Louis, Mo.
CASCADE, Peters Shoe Co., St. Louis, Mo.
CAVALIER, Peters Shoe Co., St. Louis, Mo.
CECIL, Friedman-Shelby Shoe Co., St. Louis, Mo.
CHALLENGE, Rice & Hutchins, Boston, Mass.
CHALLENGE, Endicott, Johnson & Co., Endicott, N. Y.
CHALLENGE, Banner Shoe Co., St. Louis, Mo.
CHAMPION, Dittmann Shoe Co., St. Louis, Mo.
CHAMPION, Rice & Hutchins, Boston, Mass.
CHAMPION, Peters Shoe Co., St. Louis, Mo.
CHARACTER, Joe. Rosenheim Shoe Co., Savannah, Ga.
"CHARLESTON GENTLEMAN", Payne Shoe Co., Charleston, W. Va.
"CHARLESTON LADY", Payne Shoe Co., Charlestown, W. Va.
CHATAUQUA, Bentley & Olmsted Co., Des Moines, Iowa.
CHATTER BOX, Dittmann Shoe Co., St. Louis, Mo.
CHEROKEE, Friedman-Shelby Shoe Co., St. Louis, Mo.
CHERRY BLOSSOM, Peters Shoe Co., St. Louis, Mo.
CHESTERFIELD, Ellet-Kendall Shoe Co., Kansas City, Mo.
CHIC, Friedman-Shelby Shoe Co., St. Lou's, Mo.
CHIEF, Barton Bros., Kansas City, Mo.
CHIEF ROPER, Brown Shoe Co., St. Louis, Mo.
CHILDREN'S PUMPS, H. A. Noyes, Haverhill, Mass.
CHIPPEWA DRIVER Chippewa Shoe Mfg. Co., Chippewa Falls, Wis.
CHRISTINA, Daniel Green Felt Shoe Co., Dolgeville, N. Y.
CHROME CALF, Dittmann Shoe Co., St. Louis, Mo.
CHROME X. L., Huiskamp Bros. Co., Keokuk, Iowa.
CIMARRON, Brown Shoe Co., St. Louis, Mo.
CINDERELLA, Roberts, Johnson & Rand Shoe Co., St. Louis, Mo.
CINDERELLA, Filsinger-Boette Shoe Co., St. Louis, Mo.
CINNAMON, Nolan-Earl Shoe Co., San Francisco, California.
CITY PARK, Lynchburg Shoe Co., Inc., Lynchburg, Va.
CLARICE, Thomson-Crooker Shoe Co., Boston, Mass.
CLASS "A", G. M. Kutz Shoe Co., San Francisco, California.
CLASSIC, H. Pretzfelder & Co., Baltimore, Md.
CLASSIC, Peters Shoe Co., St. Louis, Mo.
CLASSMATE, Tucker-Hagen, Chicago, Ill.
CLEOPATRA, Nathan D. Dodge Shoe Co., Newburyport, Mass.

CLIMAX, Friedman-Shelby Shoe Co., St. Louis, Mo.
CLYMER, A. Priesmeyer Shoe Co., Jefferson City, Mo.
COCK OF THE WALK, F. P. Kirkendall & Co., Omaha, Neb.
COCK OF THE WALK, Dittmann Shoe Co., St. Louis, Mo.
COLLEGE, The Kreider, Schwarz & Sallenbach Shoe Co., St. Louis, Mo.
COLLEGE GIRL, Hirth-Krause Co., Grand Rapids, Mich.
COLLEGE GIRL, Golden State Shoe Co., Los Angeles, Cal.
COLLEGE WOMAN'S, Craddock-Terry Co., Lynchburg, Va.
COLLEGIAN, E. B. Pickenbrock & Sons, Dubuque, Iowa.
COLLEGIAN, Banner Shoe Company, St. Louis, Mo.
COLONIAL, Brown Shoe Co., St. Louis, Mo.
COLONIAL CALF, Hamilton-Brown Shoe Co., St. Louis, Mo.
COLONIAL DAME, Endicott, Johnson & Co., Endicott, N. Y.
COLORADO, Dittmann Shoe Co., St. Louis, Mo.
COLUMBIA, Foot, Schulze & Co., St. Paul, Minn.
COLUMBIA, H. Pretzfelder & Co., Baltimore, Md.
COLUMBIA, Banner Shoe Company, St. Louis, Mo.
COMET, Rice & Hutchins, Boston, Mass.
COMET, Friedman-Shelby Shoe Co., St. Louis, Mo.
COMET, Hirth-Krause Co., Grand Rapids, Mich.
COMFORT, Friedman-Shelby Shoe Co., St. Louis, Mo.
COMFORT, Marion Shoe Co., Marion, Ind.
COMFORT SLIPPERS, H. Lindenberg, New Orleans, La.
COMFY, Daniel Green Felt Shoe Co., Dolgeville, N. Y.
COMPANION, Brown Shoe Co., St. Louis, Mo.
COMPASS, The Cady-Ivison Shoe Co., Cleveland, Ohio.
COMPOSITE, The Pingree Company, Detroit, Mich.
CONCHO, Friedman-Shelby Shoe Co., St. Louis, Mo.
CONCORD, Peters Shoe Co., St. Louis, Mo.
CONFIDENCE, Lynchbkurg Shoe Co., Inc., Lynchburg, Va.
CONQUEROR, Roberts, Johnson & Rand Shoe Co., St. Louis, Mo.
CONSTITUTION SCHOOL SHOE, Jos. Rosenheim Shoe Co., Savannah, Ga.
CONSOLATION, Brown Shoe Co., St. Louis, Mo.
"COPELAND," Ellis F. Copeland & Son, Brockton, Mass.
COPPER KING, Hirth-Krause Co., Grand Rapids, Mich.
CORNCUSER, Friedman-Shelby Shoe Co., St. Louis, Mo.
CORNERSTONE, Roberts, Johnson & Rand Shoe Co., St. Louis, Mo.
CORONA, Harrisburg Shoe Mfg. Co., Harrisburg, Pa.
"CORONET", Golden State Shoe Co., Los Angeles, Cal.
COTTAGE QUEEN, The Cady-Ivison Shoe Co., Cleveland, Ohio.
COUNTESS, Nathan D. Dodge Shoe Co., Newburyport, Mass.
COUNTESS POTOCKA, Ellet-Kendall Shoe Co., Kansas City, Mo.

17

COUNTRY CLUB, Hamilton-Brown Shoe Co., St. Louis, Mo.
COURT SQUARE, Carruthers-Jones Shoe Co., St. Louis, Mo.
COUTRACH, Burrow, Jones & Dyer Shoe Co., St. Louis, Mo.
COZY CORNER, Friedman-Shelby Shoe Co., St. Louis, Mo.
CRADDOCK, Craddock Terry Co., Lynchburg, Va.
CRAWFORD, C. A. Eaton Co., Brockton, Mass.
CUERO, Dittmann Shoe Co., St. Louis, Mo.
CUPID, Peters Shoe Co., St. Louis, Mo.
CURRIN, Northwestern Shoe Co., Seattle, Wash.
CUSHION, Peters Shoe Co., St. Louis, Mo.
CUSHIONET SHOES, Syracuse Shoe Mfg. Co., Syracuse, N. Y.
CUSTOM MADE, Geo. H. Snow Co., Brockton, Mass.
CUSTOM QUALITY, Geo. H. Snow Company, Brockton, Mass.
CUSTOM SHOE, Hayward Bros., Shoe Co., Omaha, Neb.
CUSTOM WELT, Dittmann Shoe Co., St. Louis, Mo.
"CUT HIDE", J. K. Orr Shoe Co., Atlanta, Ga.
CYCLONE, Dittmann Shoe Co., St. Louis, Mo.
CYCLONE, Peters Shoe Co., St. Louis, Mo.
CZARINA, Harrisburg Shoe Mfg. Co., Harrisburg, Pa.
DAHLIA, Geddes-Brown Shoe Co., Indianapolis, Ind.
DAISY, Brown Shoe Co., St. Louis, Mo.
DAISY, Friedman-Shelby Shoe Co., St. Louis, Mo.
DAISY, Peters Shoe Co., St. Louis, Mo.
DEBUTANTE, A. H. Berry Shoe Co., Portland, Maine.
DECARBURG SEAMLESS, Excelsior Shoe & Slipper Co., Cedarburg, Wis.
DEE-VEE, Diggs-Vanneman Mfg. Co., Baltimore, Md.
DEFENDER, The Cady-Ivison Shoe Co., Cleveland, Ohio.
DELINEATOR, Monadnock Shoe Co., Keene, N. H.
DELSARTE, Medlar & Holmes Co., Philadelphia, Pa.
DELTA SHOE, Delta Shoe Co., Cambridge, Mass.
DE LUXE, P. W. Minor & Son, Batavia, N. Y.
DE LYTE, A. H. Colmary & Co., Baltimore, Md.
DEPENDON, Hamilton-Brown Shoe Co., St. Louis, Mo.
DEPENDON, Dittmann Shoe Co., St. Louis, Mo.
DESIRE, Friedman-Shelby Shoe Co., St. Louis, Mo.
DE SOTO, Dittmann Shoe Co., St. Louis, Mo.
"DIAMOND QUALITY", V. Schoenecker Boot & Shoe Co., Milwaukee, Wis.
DIAMOND SPECIAL, Peters Shoe Co., St. Louis, Mo.
DIAMOND T. BRAND, J. E. Tilt Shoe Co., Chicago, Ill.
DIANA, Nathan D. Dodge Shoe Co., Newburyport, Mass.
DIPLOMA, Friedman-Shelby Shoe Co., St. Louis, Mo.
DIRECT LINE, Endicott, Johnson & Co., Endicott, N. Y.
DISTRICT 76, Noyes-Norman Shoe Co., St. Joseph, Mo.
DITTMANN'S FIRST QUALITY, Dittmann Shoe Co., St. Louis, Mo.

DITTMANN SPECIAL, Dittmann Shoe Co., St. Louis, Mo.
DIXON SCHOOL SHOE, Watson-Plummer Shoe Co., Chicago, Ill.
DOLLY GRAY, Friedman-Shelby Shoe Co., St. Louis, Mo.
DOLLY MADISON, Friedman-Shelby Shoe Co., St. Louis, Mo.
DOMINO, Brown Shoe Co., St. Louis, Mo.
DOMINO, Dittmann Shoe Co., St. Louis, Mo.
DONGOLA, Brown Shoe Co., St. Louis, Mo.
DON JUAN, Filsinger-Boette Shoe Co., St. Louis, Mo.
DORIC, A. S. Kreider Shoe Co., Anniville, Pa.
DORIS, The Kreider, Schwarz & Sallenbach Shoe Co., St. Louis, Mo.
DORIS, Thomson-Crooker Shoe Co., Boston, Mass.
DOROTHY, Roberts, Johnson & Rand Shoe Co., St. Louis, Mo.
DOROTHY, The Cady-Ivison Shoe Co., Cleveland, Ohio.
DOROTHY, Sawyer Boot & Shoe Co., Bangor, Maine.
DOROTHY DODD, Thomas G. Plant Co., Boston, Mass.
DOUBLE STANDARD, Friedman-Shelby Shoe Co., St. Louis, Mo.
DOUGLAS, W. L., W. L. Douglas Shoe Co., Brockton, Mass.
DOVE, Friedman-Shelby Shoe Co., St. Louis, Mo.
DR. BEGER, The Roth Shoe Mfg. Co., Cincinnati, Ohio.
DR. DARLING'S CUSHION SHOE. Sherwood Shoe Co., Rochester, N. Y.
DR. EDISON CUSHION SHOE, Utz & Dunn Co., Rochester, N. Y.
DR. GAUS, The Bering Shoe Co., Cincinnati, Ohio.
DR. JAEGER, C. A. Eaton Co., Brockton, Mass.
DR. LEGG'S FOOT EASE, A. M. Legg Shoe Co., Pontiac, Ills.
DR. SAWYER CUSHION WHITE HOUSE, Brown Shoe Co., St. Louis, Mo.
DR. WHEELER CUSHION, Green-Wheeler Shoe Co., Ft. Dodge, Iowa.
DRESS, Brown Shoe Co., St. Louis, Mo.
DRESSWELL, The Cady-Ivison Shoe Co., Cleveland, Ohio
DRYFUT, Geo. H. Snow Co., Brockton, Mass.
DRY SOX, F. Mayer Boot & Shoe Co., Milwaukee, Wis.
DU BARRY, Harrisburg Mfg. Co., Harrisburg, Pa.
DUCHESS, Friedman-Shelby Shoe Co., St. Louis, Mo.
"IRON CLAD", Dugan & Hudson Co., Rochester, N. Y.
DUNLAP, The K-O Shoe Co., Cincinnati, Ohio.
DUNN OF DENVER, The Jos. P. Dunn Shoe & Leather Co., Denver, Colo.
DURABLE, Dittmann Shoe Co., St. Louis, Mo.
EARL, Nolan-Earl Shoe Co., San Francisco, California.

EASE, Hamilton-Brown Shoe Co., St. Louis, Mo.
EASE, Friedman-Shelby Shoe Co., St. Louis, Mo.
"EASY", Endicott, Johnson & Co., Endicott, N. Y.
EASY, Friedman-Shelby Shoe Co., St. Louis, Mo.
EASY FEET, Peters Shoe Co., St. Louis, Mo.
EASY SHOES FOR TIRED FEET, H. Lindenberg, New
    Orleans, La.
EASY STEPPER, Peters Shoe Co., St. Louis, Mo.
"EASY STREET", J. K. Orr Shoe Co., Atlanta, Ga.
EASY WALK, Wayne Shoe Mfg. Co., Ft. Wayne, Ind.
EASY WALKER, The Cady-Ivison Shoe Co., Cleveland, Ohio.
EASY WALKER, Hamilton-Brown Shoe Co. St. Louis, Mo.
EAZ-IN-IT, Hayward Bros. Shoe Co., Omaha, Neb.
EAZE WALKER, Wise Shaw & Feder Co., Cincinnati, Ohio.
EBONY, A. Priesmeyer Shoe Co., Jefferson City, Mo.
ECHO, Brown Shoe Co., St. Louis, Mo.
ECLIPSE, Helmers Bettman, Cincinnati, Ohio.
ECLIPSE, Nathan D. Dodge Shoe Co., Newburyport, Mass.
ECONOMY, Friedman-Shelby Shoe Co., St. Louis, Mo.
ECONOMY, Dittmann Shoe Co., St. Louis, Mo.
ECONOMY, The Kreider, Schwarz & Sallenbach Shoe Co., St.
    Louis, Mo.
ECONOMY, Foot, Schulze & Co., St. Paul, Minn.
EDGEMONT SPECIAL, Jos. Rosenheim Shoe Co., Savan-
    nah, Ga.
EDITH, Friedman-Shelby Shoe Co., St. Louis, Mo.
EDUCATOR, Rice & Hutchins, Boston, Mass.
EDUCATOR SPECIAL, Rice & Hutchins, Boston, Mass.
"EDWIN CLAPP", Edwin Clapp & Son, Inc., East Weymouth,
    Mass.
ELBERTA PEACH, Jos. Rosenheim Shoe Co., Savannah, Ga.
ELECTRA, A. Priesmeyer Shoe Co., Jefferson City, Mo.
ELECTRA, Nathan D. Dodge Shoe Co., Newburyport, Mass.
ELEGANT GENTLEMEN, Ellet-Kendall Shoe Co., Kansas
    City, Mo.
ELIOT SCHOOL SHOE, Endicott, Johnson & Co., Endicott,
    New York.
ELITE, Dittmann Shoe Co., St. Louis, Mo.
ELIZABETH, Nathan D. Dodge Shoe Co., Newburyport, Mass.
ELKIDE, Friedman-Shelby Shoe Co., St. Louis, Mo.
ELK LODGE, Burrow, Jones & Dyer Shoe Co., St. Louis, Mo.
ELKSKIN, Rice & Hutchins, Boston, Mass.
ELK SKIN, Herald Bersck Shoe Co., Grand Rapids, Mich.
ELKSOLE, Rice & Hutchins, Boston, Mass.
EL RENO, Dittmann Shoe Co., St. Louis, Mo.
ELSIE'S BEST, The Cady-Ivison Shoe Co., Cleveland, Ohio.
ELZEY R., Roberts, Johnson & Rand Shoe Co., St. Louis, Mo.
EMERSON, Emerson Shoe Co., Rockland, Mass.

EMPEROR, Marion Shoe Co., Marion, Ind.
EMPIRE, Herold Bertsck Shoe Co., Grand Rapids, Mich.
EMPRESS, Nathan D. Dodge Shoe Co., Newburyport, Mass.
EMPRESS, Peters Shoe Co., St. Louis, Mo.
ENDURANCE, Endicott, Johnson & Co., Endicott, New York.
ENDURANCE, Hamilton-Brown Shoe Co., St. Louis, Mo.
ENDURANCE, Friedman-Shelby Shoe Co., St. Louis, Mo.
ENDWELL, Endicott, Johnson & Co., Endicott, N. Y.
ENERGY, Hamilton-Brown Shoe Co., St. Louis, Mo.
ENERGY, Endicott, Johnson & Co., Endicott, N. Y.
ENTERPRISE, Brown Shoe Co., St. Louis, Mo.
EQUALITY, A. Priesmeyer Shoe Co., Jefferson City, Mo.
EQUIPOISE, Slater & Morrill, Inc., South Braintree, Mass.
ERICA, Rice & Hutchins, Boston, Mass.
ESSEX, Golden State Shoe Co., Los Angeles, California.
ESTELLE, Roberts, Johnson & Rand Shoe Co., St. Louis, Mo.
EUREKA, Daniel Greenfelt Shoe Co., Dolgeville, N. Y.
EVANGELINE, A. H. Berry Shoe Co., Portland, Maine.
EVENTRED, Dugan & Hudson Co., Rochester, N. Y.
EVER WEAR SHOE, Xenia Shoe Mfg. Co., Xenia, Ohio
EVERYBOY, Endicott, Johnson & Co., Endicott, N. Y.
EXCELLO, The Kreider, Schwarz & Sallenbach Shoe Co., St.
    Louis, Mo.
EXCELLO, A. S. Kreider Shoe Co., Annville, Pa.
EXCELSIOR, Excelsior Shoe & Slipper Co., Cedarburg, Wis.
EXCELSIOR, Excelsior Shoe Co., Portsmouth, Ohio.
EXCELSIOR, Rice & Hutchins, Boston, Mass.
EXPERT SHOE, Thomas Shoe Co., Charleston, W. Va.
EXPOSITION, Brown Shoe Co., St. Louis, Mo.
EXPRESS, Friedman-Shelby Shoe Co., St. Louis, Mo.
"E. C. SKUFFER", Payne Shoe Co., Charleston, W. Va.
E. C. SKUFFERS, E. J. Egan & Company, San Francisco,
    Cal.
E. C. SKUFFERS, Engle-Cone Shoe Co., Boston, Mass.
E. C. VENTELATED SHOE, Engel-Cone Shoe Co., Boston,
    Mass.
"E. J. CO.", Endicott, Johnson & Co., Endicott, N. Y.
E-Z, Hayward Bros. Shoe Co., Omaha, Nebr.
E. Z. CUSHION, John Kelly, Inc., Rochester, N. Y.
E. Z. CUSHION, The Bering Shoe Co., Cincinnati, Ohio.
"E. Z." SHOE, A. F. Smith Shoe Co., Lynn, Mass.
"E. Z." WALKER, F. P. Kirkendall & Co., Omaha, Nebr.
E-Z WALKER, Peters Shoe Co., St. Louis, Mo.
FAIR DEAL, Peters Shoe Co., St. Louis, Mo.
FAIR MAIDEN, Bentley & Olmsted Co., Des Moines, Iowa
FAIR PLAY, Brown Shoe Co., St. Louis, Mo.
FAIRY QUEEN, Filsinger-Boette Shoe Co., St. Louis, Mo
FAIRY, Peters Shoe Co., St. Louis, Mo.

FAITH, Hamilton-Brown Shoe Co., St. Louis, Mo.
FALCON, A. Priesmeyer Shoe Co., Jefferson City, Mo.
FARRELL, Hamilton-Brown Shoe Co., St. Louis, Mo.
FASHION, Dougherty-Fithian Shoe Co., Portland, Oregon.
FASHION LEADER, Lynchburg Shoe Co., Inc., Lynchburg, Va.
FASHION, Golden State Shoe Co., Los Angeles, Cal.
FASHIONABLE, Brown Shoe Co., St. Louis, Mo.
FAST TIMES, Peters Shoe Co., St. Louis, Mo.
FAT ANKLE, Roberts, Johnson & Rand Shoe Co., St. Louis, Mo.
FAT BABY, Friedman-Shelby Shoe Co., St. Louis, Mo.
FAT BABY, Peters Shoe Co., St. Louis, Mo.
FAT BABY, Hamilton-Brown Shoe Co., St. Louis, Mo.
FAVORITE, Friedman-Shelby Shoe Co., St. Louis, Mo.
FEDERAL, A. Priesmeyer Shoe Co., Jefferson City, Mo.
FELLOWCRAFT, Churchill & Alden Co., also Ralston Health Shoemakers, Campello, Mass.
FERNDALE, Peters Shoe Co., St. Louis, Mo.
FERRIS SUPERB, I. Ferris & Co., Camden, N. J.
FIFTH AVENUE, The Cady-Ivison Shoe Co., Cleveland, Ohio.
FIFTH AVENUE, Ellet-Kendall Shoe Co., Kansas City, Mo.
FIGARO, Carruthers-Jones Shoe Co., St. Louis, Mo.
FIRESIDE, Peters Shoe Co., St. Louis, Mo.
FIRFELT, Worcester Slipper Co., Worcester, Mass.
FIRM FOUNDATION, Brown Shoe Co., St. Louis, Mo.
FIRST CONSUL, Milford Shoe Co., Milford, Mass.
FIRST STEP, Rice & Hutchins, Boston, Mass.
FIT EASY, C. P. Ford & Co., Rochester, N. Y.
FIVETOE, Wolfe Bros. Shoe Co., Columbus, Ohio.
FLEETFOOT, Elk Skin Moccasin Mfg. Co., Ypsilanti, Mich.
FLEXEZE, P. W. Minor & Son, Batavia, N. Y.
FLEXIBLE, Whittinghill-Harlow Shoe Co., St. Joseph, Mo.
FLEX-WELT, The Rich Shoe Co., Milwaukee, Wis.
FLINTSTONE, Hayward Bros. Shoe Co., Omaha, Neb.
FLORETTE, Daniel Green Felt Shoe Co., Dolgeville, N. Y.
FLORIO, Daniel Green Felt Shoe Co., Dolgeville, N. Y.
FLORODORA, The Cady-Ivison Shoe Co., Cleveland, Ohio.
FLORSHIEM, Florshiem Shoe Co., Chicago, Ill.
FLYER, Endicott, Johnson & Co., Endicott, N. Y.
FOOT CARESSER, Jos. Rosenheim Shoe Co., Savannah, Ga.
FOOT CULTURE, Roberts, Johnson & Rand Shoe Co., St. Louis, Mo.
FOOT DOCTOR, Heywood Boot & Shoe Co., Worcester, Mass.
FOOT FORM, Peters Shoe Co., St. Louis, Mo.

FOOT ROOM, Brown Shoe Co., St. Louis, Mo.
FOOT SCHULZE, Foot, Schulze & Co., St. Paul, Minn.
FORBUSH, Forbush Shoe Co., North Grafton, Mass.
FOREMOST, Churchill & Alden Co., also Ralston Health
   Shoemakers, Campello, Mass.
FORESTER, Noyes-Norman Shoe Co., St. Joseph, Mo.
FOSTER, Jno. F. Foster Co., Beloit, Wis.
FRANKLIN SHOE, Franklin Shoe Co., Boston, Mass.
FREE SILVER, Friedman-Shelby Shoe Co., St. Louis, Mo.
FRENCH SHOE, J. E. French & Co., Rockland, Mass.
FRISCO, Dittmann Shoe Co., St. Louis, Mo.
FROLIC, Brown Shoe Co., St. Louis, Mo.
FRONT RANK, Peters Shoe Co., St. Louis, Mo.
FRONTIER, Dittmann Shoe Co., St. Louis, Mo.
FULL DRESS, The Apex Shoe Factory, 300 Decatur St., New
   Orleans, La.
FULTON, Endicott, Johnson & Co., Endicott, N. Y.
FUN, Brown Shoe Co., St. Louis, Mo.
FURNITURE CITY GIRL, Hirth-Krause Co., Grand Rapids,
   Mich.
FUSSY, Friedman-Shelby Shoe Co., St. Louis, Mo.
GARDEN, Friedman-Shelby Shoe Co., St. Louis, Mo.
GASTON, Friedman-Shelby Shoe Co., St. Louis, Mo.
GENTILITY, Wertheimer-Swarts Shoe Co., St. Louis, Mo.
GERMAN, Brown Shoe Co., St. Louis, Mo.
GIANT CALF, Roberts, Johnson & Rand Shoe Co., St. Louis,
   Mo.
GIBRALTAR, Hayward Bros. Shoe Co., Omaha, Neb.
GIBRALTER, Dittmann Shoe Co., St. Louis, Mo.
GIPSY, Peters Shoe Co., St. Louis, Mo.
GIRL OF TODAY, The Cady-Ivison Shoe Co., Cleveland, O.
GITCHE-GAMER, Northern Shoe Co., Duluth, Minn.
GLADSTONE, Hamburger Bros. Shoe Co., St. Louis, Mo.
GLENDALE, Worcester Slipper Co., Worcester, Mass.
GLEN MARY, Carruthers-Jones Shoe Co., St. Louis, Mo.
GLENWOOD, Peters Shoe Co., St. Louis, Mo.
GLORIA, The Pingree Company, Detroit, Mich.
GLORY, Roberts, Johnson & Rand Shoe Co., St. Louis, Mo.
GLOVE ELK MOUND CITY, Brown Shoe Co., St. Louis, Mo.
GLOVE ELK OUTING, Brown Shoe Co., St. Louis, Mo.
GLOVER MAID, Wertheimer-Swarts Shoe Co., St. Louis, Mo.
GOKEY, Wm N. Gokey Shoe Co., Jamestown, N. Y.
GOLD BOND, Roberts, Johnson & Rand Shoe Co., St. Louis,
   Mo.
GOLD DUST TWINS, Jos. Rosenheim Shoe Co., Savannah,
   Mo.
GOLD DUST TWINS, Jos. Rosenheim Shoe Co., Savannah,
   Ga.

23

GOLDEN, Nolan-Earl Shoe Co., San Francisco, California.
GOLDEN, Golden Sporting Shoe Co., Brockton, Mass.
GOLDEN RULE, McCord-Donovan Shoe Co., St. Joseph, M
GOLF, Marion Shoe Co., Marion, Ind.
GOLFER, Burrow, Jones & Dyer Shoe Co., St. Louis, Mo.
GOLIAD, Friedman-Shelby Shoe Co., St. Louis, Mo.
GOMPERS, Filsinger-Boette Shoe Co., St. Louis, Mo.
GOOD AS WHEAT, Friedman-Shelby Shoe Co., St. Louis, M
GOOD BOYS, Barker, Brown & Compa, Huntington, Ind.
GOOD FEELER, Peters Shoe Co., St. Louis, Mo.
GOOD FOR BAD BOYS, Barker, Brown & Compa, Hunting
    ton, Ind.
GOODSELLER, Jos. Rosenheim Shoe. Co., Savannah, Ga.
GOOD VALUE, Peters Shoe Co., St. Louis, Mo.
GORILLA SHOES, H. H. Brown Co., Boston, Mass.
GOT 'EM, Peters Shoe Co., St. Louis, Mo.
GOT THERE, Filsinger-Boette Shoe Co., St. Louis, Mo.
GOVERNOR, The Pingree Company, Detroit, Mich.
GRAMMAR SCHOOL, The Jos. P. Dunn Shoe & Leather Co.
    Denver, Colo.
GRAND, Bode-Larson Shoe Co., Keokuk, Iowa.
GRANT, Hamburger Bros. Shoe Co., St. Louis, Mo.
GRAY BROTHERS, Syracuse Shoe Mfg. Co., Syracuse, N. Y.
GREAT, Friedman-Shelby Shoe Co., St. Louis, Mo.
GREATEST, Roberts, Johnson & Rand Shoe Co., St. Louis.
    Mo.
GREAT REPUBLIC, Brown Shoe Co., St. Louis, Mo.
GREAT SCOTT, Peters Shoe Co., St. Louis, Mo.
GRECIAN, Brown Shoe Co., St. Louis, Mo.
GREEN-WHEELER, Green-Wheeler Shoe Co., Ft. Dodge,
    Iowa.
GREENLAND, Brown Shoe Co., St. Louis, Mo.
GREELEY, A. W. Greeley, Haverhill, Mass.
GRIFFINS WONSEAM, W. H. Griffin, Manchester, N. H.
GUN, Friedman-Shelby Shoe Co., St. Louis, Mo.
GUN METAL, Herold Bertsck Shoe Co., Grand Rapids, Mich.
GYPSY, Brown Shoe Co., St. Louis, Mo.
GYPSY, Roberts, Johnson & Rand Shoe Co., St. Louis, Mo.
HAMMERER, Carruthers-Jones Shoe Co., St. Louis, Mo.
HAMILTON, Hamilton-Brown Shoe Co., St. Louis, Mo.
HANAN, Hanan & Son, Brooklyn, Mass.
HAND CRAFT, Thompson Bros., Campello, Mass.
"HAND-OVER, Payne Shoe Co., Charleston, W. Va. •
HANNAH, Nathan D. Dodge Shoe Co., Newburyport, Mass.
HANSE, Watson Plummer Shoe Co., Chicago, Ill.
HAPPY, Friedman-Shelby Shoe Co., St. Louis, Mo.
HAPPY DAYS, Chipman, Harwood & Co., Boston, Mass.
HAPPY DAY, Brown Shoe Co., St. Louis, Mo.

HAPPY DAYS, Peters Shoe Co., St. Louis, Mo.
HARD HITTER, Dittmann Shoe Co., St. Louis, Mo.
HARD KNOCKER, F. P. Kirkendall & Co., Omaha, Neb.
HARD KNOCKS, Rice & Hutchins, Boston, Mass.
HARDPAN, Friedman-Shelby Shoe Co., St. Louis, Mo.
HARD PAN, Marion Shoe Co., Marion, Ind.
HARD PAN, Herald Bertsck Shoe Co., Grand Rapids, Mich.
HARDWARE, Hamilton-Brown Shoe Co., St. Louis, Mo.
HARDWEAR, Hirth-Krause Co., Grand Rapids, Mich.
HARLOW, McCord-Donovan Shoe Co., St. Joseph, Mo.
HARVARD, Endicott- Johnson & Co., Endicott, N. Y.
HARVARD, Dougherty-Fithian Shoe Co., Portland, Oregon.
HARVEST, Roberts, Johnson & Rand Shoe Co., St. Louis,
    Mo.
HARVEY W., Roberts, Johnson & Rand Shoe Co., St. Louis,
    Mo.
HAWTHORNE, Pingree Company, The, Detroit. Mich.
H. C. WOOD, Roberts, Johnson & Rand Shoe Co., St. Louis,
    Mo.
HELEN HUNT, Jos. Rosenheim Shoe Co., Savannah, Ga.
HERO, Peters Shoe Co., St. Louis, Mo.
HERO, Friedman-Shelby Shoe Co., St. Louis, Mo.
HERRICK, The G. W. Herrick Shoe Co., Lynn, Mass.
HEYWOOD SPECIAL, Heywood Boot & Shoe Co., Wor-
    cester, Mass.
HIAWATHA MOCCASIN, Sawyer Boot & Shoe Co., Bangor,
    Maine.
HICKORY, Roberts, Johnson & Rand Shoe Co., St. Louis, Mo.
HIGH ART, Banner Shoe Company, St. Louis, Mo.
HIGHBINDER, Guthmann, Carpenter & Telling, Chicago, Ill.
HIGH CUTS, Barker, Brown & Co., Huntington, Ind.
HIGH FLYER, Calm, Nickelsburg & Co., San Francisco, Cal.
HIGH GRADE, Peters Shoe Co., St. Louis, Mo.
HIGHLAND KID, Hamilton-Brown Shoe Co., St. Louis, Mo.
HIGH LIFE, The Cady-Ivison Shoe Co., Cleveland, Ohio.
HIGH LIFE, Morse & Rogers, New York, N. Y.
HIGH QUALITY, Hayward Bros. Shoe Co., Omaha, Neb.
HI NOTCH, A. Priesmeyer Shoe Co., Jefferson City, Mo.
HIP-O-TAN, Calm, Nickelsburg & Co., San Francisco, Cal.
HIS EXCELLENCY, Jos. Rosenheim Shoe Co., Savannah, Ga.
HIS HIGHNESS, I. P. Farnum, Chicago, Ill.
HOCKER BOOTEE, The Manss Shoe Mfg. Co., Cincinnati,
    Ohio.
HOCKEY, Endicott, Johnson & Co., Endicott, N. Y.
HODGDON, The Cady-Ivison Shoe Co., Cleveland, Ohio.
HOLD-SHAPE, Wise, Shaw & Feder Co., Cincinnati, Ohio.
HOME, Friedman-Shelby Shoe Co., St. Louis, Mo.
HOME GROWN, Wertheimer-Swarts Shoe Co., St. Louis, Mo.

HOME MADE FOR WEAR, Wertheimer-Swarts Shoe Co., St. Louis, Mo.
HOME RUN, A. Priesmeyer Shoe Co., Jefferson, Mo.
HOMER SHOE, The Homer Shoe Co., Oxford, Mass.
HOMESTAKE, Barton, Bros., Kansas City, Mo.
HONESTY, Friedman-Shelby Shoe Co., St. Louis, Mo.
HONEST QUALITY, Nolan-Earl Shoe Co., San Francisco, California.
HONORBILT, The Washington Shoe Mfg. Co., Seattle, Wash.
HONORBILT, F. Mayer Boot & Shoe Co., Milwaukee, Wis.
HONOR BOUND, Lynchburg Shoe Co., Inc., Lynchburg, Va.
HONOR ROLL, Arnold Henegar Dyle Co., Knoxville, Tenn.
HOOSIER SCHOOL SHOES, The Tappan Shoe Mfg. Co., Goldwater, Mich.
HOPE, Hamilton-Brown Shoe Co., St. Louis, Mo.
HORNET, Friedman-Shelby Shoe Co., St. Louis, Mo.
HOT SHOT, Peters Shoe Co., St. Louis, Mo.
HOUSEHOLD, Peters Shoe Co., St. Louis, Mo.
HOUSEHOLD, Hamilton-Brown Shoe Co., St. Louis, Mo.
HOUSE SLIPPERS, H. Lindenberg, New Orleans, La.
HUB, Jos. Rosenheim Shoe Co., Savannah, Ga.
HUB, Hamilton-Brown Shoe Co., St. Louis, Mo.
HUB LINE, Endicott, Johnson & Co., Endicott, N. Y.
HUMANIC, Nesmith Shoe Co., Brockton, Mass.
HUMMER, Friedman-Shelby Shoe Co., St. Louis, Mo.
HUNTER, Hamilton-Brown Shoe Co., St. Louis, Mo.
HUNTING BOOTS, Elk Skin Moccasin Mfg. Co., Ypsilanti, Mich.
HURLEY, Hurley Shoe Co., Rockland, Mass.
HUSSCO, Honesdale Union-Stamp Shoe Co., Honesdale, Pa.
HYKLASS, Geo. H. Snow Co., Brockton, Mass.
I-AM-A-FIT, Hayward Bros. Shoe Co., Omaha, Neb.
ICE KING, C. W. Johnson, Natick, Mass.
IDA, Brown Shoe Co., St. Louis, Mo.
IDAHO, Dittmann Shoe Co., St. Louis, Mo.
IDEAL, Brown Shoe Co., St. Louis, Mo.
IDEAL, Peters Shoe Co., St. Louis, Mo.
IMPERIAL, The Bering Shoe Co., Cincinnati, Ohio.
IMPERIAL, A. Preismeyer Shoe Co., Jefferson City, Mo.
IMPERIAL, Friedman-Shelby Shoe Co., St. Louis, Mo.
INDEPENDENT, Northern Shoe Co., Duluth, Minn.
INDEX KID ARCH-DROP, Chas. Case Shoe Co., Worcester, Mass.
INDIAN TERRITORY, Dittmann Shoe Co., St. Louis, Mo.
INSPECTOR, Rice & Hutchins, Boston, Mass.
INSURGENT, Endicott, Johnson & Co., Endicott, N. Y.
INTEGRITY, Nolan-Earl Shoe Co., San Francisco, Cal.
INVINCIBLE, Preston B. Keith Co., Brockton, Mass.

IOWA, Dittmann Shoe Co., St. Louis, Mo.
IOWA, Hamilton-Brown Shoe Co., St. Louis, Mo.
IRENE, Roberts, Johnson & Rand Shoe Co., St. Louis, Mo.
IRON CLAD, Dougherty-Fithian Shoe Co., Portland, Oregon.
IRONSIDES, Jos. Rosenheim Shoe Co., Savannah, Ga.
ISABELLE, Endicott, Johnson & Co., Endicott, N.Y.
J. & M., Johnson & Murphy, Newark, N. J.
J. & K., Julian-Kokenge Shoe Co., Cincinnati, Ohio.
J. C. ROBERTS, Roberts, Johnson & Rand Shoe Co., St. Louis, Mo.
JACKSON BOOTEE, Roberts, Johnson & Rand Shoe Co., St. Louis, Mo.
JAMES MEANS, C. A. Eaton Co., Brockton, Mass.
JANET, Daniel Green Felt Shoe Co., Dolgeville, N. Y.
JEFF, Peters Shoe Co., St. Louis, Mo.
JEWEL, Roberts, Johnson & Rand Shoe Co., St. Louis, Mo.
JEWELL, Brown Shoe Co., St. Louis, Mo.
JIMMY BRITT, Dittmann Shoe Co., St. Louis, Mo.
JOHANSEN SHOE, Johansen Bros. Shoe Co., St. Louis, Mo.
JOHN MITCHELL, M. A. Packard Co., Brockton, Mass.
JOHN TELLING, Guthmann, Carpenter & Telling, Chicago, Ill.
JOHNS HOPKINS, H. Pretzfelder & Co., Baltimore, Md.
JOHNNY, Friedman-Shelby Shoe Co., St. Louis, Mo.
JOSLIN, The Homer Shoe Co., Oxford, Mass.
JOYFUL, Friedman-Shelby Shoe Co., St. Louis, Mo.
JUBILEE, A. Priesmeyer Shoe Co., Jefferson City, Mo.
JULIA MARLOWE, The Rich Shoe Co., Milwaukee, Wis.
JUPITER, Hirth-Krause Co., Grand Rapids, Mich.
JUST RIGHT, Harrisburg Shoe Mfg. Co., Harrisburg, Pa.
JUST WRIGHT, E. T. Wright & Co., Inc., Rockland, Mass.
K. K. K. K., Ellet-Kendall Shoe Co., St. Louis, Mo.
K-O SHOE, The K-O Shoe Co., Cincinnati, Ohio.
K. & T., Brown Shoe Co., St. Louis, Mo.
K-Z, Kalt Zimmers Mfg. Co., Milwaukee, Wis.
K-Z TREDSHURE, Kalt Zimmers Mfg. Co., Milwaukee, Wis.
"KANAWHA", Payne Shoe Co., Charleston, W. Va.
KANTBEBEAT, Banner Shoe Company, St. Louis, Mo.
KANT-WEAR-OUT, Excelsior Shoe Co., Portsmouth, Ohio.
KATY, Endicott, Johnson & Co., Endicott, N. Y.
KEEP COOL, Brown Shoe Co., St. Louis, Mo.
KEITH, Preston B. Keith Co., Brockton, Mass.
KEITH'S KONQUEROR, Preston B. Keith Co., Brockton, Mass.
KENOZA, Kenoza Shoe Co., Haverhill, Mass.
"KESCO", The Kepner-Scott Shoe Co., Orwigsburg, Pa.
KEYNOTE, Brown Shoe Co., St. Louis, Mo.
KICKPROOF, The Tappan Shoe Mfg. Co., Coldwater, Mich.

KILEY SHOE, The Manss Shoe Mfg. Co., Cincinnati, Ohio.
"KIMLAND SHOE",. Williams Kimland Co., South Braintree,
. Mass.
KIMONA, Star Baby Shoe Co., Minneapolis, Minn.
KINDERGARTEN, J. P. Hartray Shoe Co., Chicago, Ill.
KING, Roberts, Johnson & Rand Shoe Co., St. Louis, Mo.
"KING BEE", J. K. Orr Shoe Co., Atlanta, Ga.
KING QUALITY, Arnold Shoe Co., North Abington, Mass.
KING PHILIP, W. N. May, Bridgewater, Mass.
KITTY-KAT, James Clark Leather Co., St. Louis, Mo.
KLONDIKE, H. D. Raff & Co., Chicago, Ill.
KNOCKABOUT, Geddes-Brown Shoe Co., Indianapolis, Ind.
KNOCK-A-BOUT, Williams, Hoyt & Co., Rochester, N. Y.
KNOCKAROUND, Endicott, Johnson & Co., Endicott, N. Y.
KNOX CUSHION SOLE, Dougherty-Fithian Shoe Co., Port-
land, Oregon.
KOLD KILLER, Peters Shoe Co., St. Louis, Mo.
KORRECTOW, Geddes-Brown Shoe Co., Indianapolis, Ind.
KOZY, Kozy Slipper Co., Lynn, Mass.
KREEP A WA, Blum Shoe Mfg. Co., Danville, Ky.
KREIDER'S KICKER, The Kreider, Schwarz & Sallenbach
Shoe Co., St. Louis, Mo.
KREIDER'S KICKERS, A. S. Kreider Shoe Co., Annville, Pa.
KROM CALF, Friedman-Shelby Shoe Co., St. Louis, Mo.
KROMELK, Endicott, Johnson & Co., Endicott, N. Y.
KUTZ SHOE, G. M. KUTZ SHOE CO., San Francisco, Cal.
LADY LEE, Carruthers-Jones Shoe Co., St. Louis, Mo.
LADY-LIKE, E. B. Pickenbrock & Sons, Dubuque, Iowa.
LADY JEFFERSON, Burrow, Jones & Dyer Shoe Co., St.
Louis, Mo.
LADY WALKER, John Kelly, Inc., Rochester, N. Y.
LAKE SHORE LINE, The Cady-Ivison Shoe Co., Cleveland,
Ohio.
LANCREATE, O'Sullivan Bros. Co., Lowell, Mass.
LAUREL, Rice & Hutchins, Boston, Mass.
LAUREL, Peters Shoe Co., St. Louis, Mo.
LAUREL, A. Priesmeyer Shoe Co., Jefferson City, Mo.
LEADER, Friedman-Shelby Shoe Co., St. Louis, Mo.
LEADING LADY, F. Mayer Boot & Shoe Co., Milwaukee,
Wis.
LESTERSHIRE, Endicott, Johnson & Co., Endicott, N. Y.
LEVIE. Levie Shoe Co., Chicago, Ill.
LEXINGTON, Craddock-Terry Co., Lynchburg, Va.
LIBERATOR, The Tappan Shoe Mfg. Co., Coldwater, Mich.
LIBRARY, Brown Shoe Co., St. Louis, Mo.
LIFE, Hamilton-Brown Shoe Co., St. Louis, Mo.
LILAC, Brown Shoe Co., St. Louis, Mo.
LILY, Brown Shoe Co., St. Louis, Mo.

LIMIT, Brown Shoe Co., St. Louis, Mo.
LINCOLN, Golden State Shoe Co., Los Angeles, California.
LINCOLN, H. Pretzfelder & Co., Baltimore, Md.
LION BRAND, Harsh & Edmonds Shoe Co., Milwaukee, Wis.
LION-TAN, Nolan-Earl Shoe Co., San Francisco, California.
LITTLE BABY KING, Mrs. A. R. King, Corp., Lynn, Mass.
LITTLE BROTHER, Friedman-Shelby Shoe Co., St. Louis, Mo.
"LITTLE CHUM", L. B. Evans Son Co., Wakefield, Mass.
LITTLE DAHLIA, Geddes-Brown Shoe Co., Indianapolis, Ind.
LITTLE KING, Peters Shoe Co., St. Louis, Mo.
LITTLE LADY KING, Mrs. A. R. King, Lynn, Mass.
LITTLE MISS KING, Mrs. A. R. King Corp., Lynn, Mass.
LITTLE RED SCHOOL HOUSE, Watson Plummer Shoe Co., Chicago, Ill.
LITTLE RICHIE, Jos. Rosenheim Shoe Co., Savannah, Ga.
LITTLE SOLDIER, Wolfe Shoe Mfg. Co., Allentown, Pa.
LITTLE SWEETHEART, A. Priesmeyer Shoe Co., Jefferson City, Mo.
LITTLE TRAMP, The Cady-Ivison Shoe Co., Cleveland, Ohio.
LITTLE WANDERER, The Utz & Dunn Co., Rochester, N. Y.
LIVE OAK, Peters Shoe Co., St. Louis, Mo.
LIVE WIRE, Hamilton-Brown Shoe Co., St. Louis, Mo.
LOCUST POST, Arnold-Henegar-Doyle Co., Knoxville, Tenn.
LOLA, Friedman-Shelby Shoe Co., St. Louis, Mo.
LONG BRANCH, Friedman-Shelby Shoe Co., St. Louis, Mo.
LONG WEARER, Harrisburg Shoe Mfg. Co., Harrisburg, Pa.
LONGWORTH, Brown Shoe Co., St. Louis, Mo.
LUCILLE, Dittmann Shoe Co., St. Louis, Mo.
LUCKY, More & Rogers, New York, N. Y.
LUCKY STRIKE, Endicott, Johnson & Co., Endicott, N. Y.
LUCY LOCKHART, Jos. Rosenheim Shoe Co., Savannah, Ga.
LUDGATE, J. E. French & Co., Rockland, Mass.
LUXURY, Friedman-Shelby Shoe Co., St. Louis, Mo.
LUZON, Nolan-Earl Shoe Co., San Francisco, California.
M. & K., Macdonald & Kiley Co., Cincinnati, Ohio.
MABEL, Friedman-Shelby Shoe Co., St. Louis, Mo.
MADE FOR WEAR, American Hand Sewed Shoe Co., Omaha, Neb.
MADISON, Friedman-Shelby Shoe Co., St. Louis, Mo.
MAGNET, Friedman-Shelby Shoe Co., St. Louis, Mo.
MAGNET, The Rich Shoe Co., Milwaukee, Wis.
MAGNET, Roberts, Johnson & Rand Shoe Co., St. Louis, Mo.
MAGNOLIA, Rice & Hutchins, Boston, Mass.
MAGNOLIA, Peters Shoe Co., St. Louis, Mo.
MAID OF HONOR, The Cady-Ivison Shoe Co., Cleveland, O.
MAIDWELL, Foot, Schulze & Co., St. Paul, Minn.

MAMIE, Friedman-Shelby Shoe Co., St. Louis, Mo.
MANHATTAN, Banner Shoe Company, St. Louis, Mo.
MANSION, Brown Shoe Co., St. Louis, Mo.
MANSPROOF, Hamilton-Brown Shoe Co., St. Louis, Mo.
MANSO, The Manso Shoe Mfg. Co., Cincinnati, Ohio.
MANSO URFIT, The Manso Shoe Mfg. Co., Cincinnati, Ohio.
MARATHON, Peters Shoe Co., St. Louis, Mo.
MARCOE, Marion Shoe Co., Marion, Ind.
MARS, Hirth-Krause Co., Grand Rapids, Mich.
MARSH COMFORT, The Henry C. Werner Co., Columbus, Ohio.
MARSHALL, C. S. Marshall Co., Brockton, Mass.
MARSTON SHOE, THE, Packard, Marston & Brooks, Inc., Danvers, Mass.
MARTHA, Peters Shoe Co., St. Louis, Mo.
MARTHA WASHINGTON, F. Mayer Boot & Shoe Co., Milwaukee, Wis.
MARVEL, Rice & Hutchins, Boston, Mass.
MARVELLO, A. S. Kreider Shoe Co., Annville, Pa.
MARVELLO, The Kreider, Schwarz & Sallenbach Shoe Co., St. Louis, Mo.
MARY, Peters Shoe Co., St. Louis, Mo.
MARY LYNN, Arnold Henegar Doyle Co., Knoxville, Tenn.
MARYLAND, Friedman-Shelby Shoe Co., St. Louis, Mo.
MASTER, Hamilton-Brown Shoe Co., St. Louis, Mo.
MASTERBILT, Burrow, Jones & Dyer Shoe Co., St. Louis, Mo.
MASTERPIECE, Jos Rosenheim Shoe Co., Savannah, Ga.
MATCHLESS, Matchless Shoe Co., Boston, Mass.
MATCH-US, Wertheimer-Swarts Shoe Co., St. Louis, Mo.
MAY, Friedman-Shelby Shoe Co., St. Louis, Mo.
MAY DAY, Roberts, Johnson & Rand Shoe Co., St. Louis, Mo.
MAY DAY, Friedman-Shelby Shoe Co., St. Louis, Mo.
MAYFAIR, Rice & Hutchins, Boston, Mass.
MAYFLOWER, Roberts, Johnson & Rand Shoe Co., St. Louis, Mo.
MAY MANTON, The Roth Shoe Mfg. Co., Cincinnati, Ohio.
MAYNARD, Maynard Shoe Company, Claremont, N. H.
MAY QUEEN, A. Priesmeyer Shoe Co., Jefferson City, Mo.
MAZEPPA, Roberts, Johnson & Rand Shoe Co., St. Louis, Mo.
MENZ "EASE", Menzies Shoe Co., Detroit, Mich.
MERCANTILE, Burrow, Jones & Dyer Shoe Co., St. Louis, Mo.
MERCURY, More & Rogers, New York, N. Y.
MERITUS, Guthmann, Carpenter & Telling, Chicago, Ill.
MERRY MACK, Jos. Rosenheim Shoe Co., Savannah, Ga.
MERRY WIDOW, The Cady-Ivison Shoe Co., Cleveland, O.
MESSENGER, Filsinger-Boette Shoe Co., St. Louis, Mo.

METROPITAN, J. J. Lattemann Shoe Mfg. Co., Brooklyn, N. Y.
MIDDLESEX, Rice & Hutchins, Boston, Mass.
MIKADO, Dittmann Shoe Co., St. Louis, Mo.
MILADY, Dittmann Shoe Co., St. Louis, Mo.
MILEAGE, Endicott, Johnson & Co., Endicott, N . Y.
MINEPROOF, Hamilton-Brown Shoe Co., St. Louis, Mo.
MINER'S, Brown Shoe Co., St. Louis, Mo.
MINOR'S EASY SHOE, P. W. Minor & Son, Batavia, N. Y.
MINOR'S SENSIBLE SHOE, P. W. Minor & Son, Batavia, N. Y.
MISS CHICAGO, J. P. Hartray Shoe Co., Chicago, Ill.
MISSION, F. P. Kirkendall & Co., Omaha, Neb.
MISS MYRTLE, Peters Shoe Co., St. Louis, Mo.
MISSOURI, Dittmann Shoe Co., St. Louis, Mo.
MOBILE, Friedman-Shelby Shoe Co., St. Louis, Mo.
MOCCASIN SHEEP, Elk Skin Moccasin Mfg. Co., Ypsilanti, Mich.
MODERN, Friedman-Shelby Shoe Co., St. Louis, Mo.
MOJAVE, Friedman-Shelby Shoe Co., St. Louis, Mo.
MONARCH, Friedman-Shelby Shoe Co., St. Louis, Mo.
MONARCH, Rice & Hutchins, Boston, Mass.
MONEY-BAK, Crowder-Cooper Shoe Co., Indianapolis, Ind.
MOTOR, Friedman-Shelby Shoe Co., St. Louis, Mo.
MOULDER, Endicott, Johnson & Co., Endicott, N. Y.
MOULDERS, Brown Shoe Co., St. Louis, Mo.
MOUND CITY ARMY, Brown Shoe Co., St. Louis, Mo.
MOUND CITY CHIEF, Brown Shoe Co., St. Louis, Mo.
MOUND CITY EASY, Brown Shoe Co., St. Louis, Mo.
MOUND CITY LEADER, Brown Shoe Co., St. Louis, Mo.
MOUND CITY OVERLAND, Brown Shoe Co., St. Louis, Mo
MOUND CITY WADER, Brown Shoe Co., St. Louis, Mo.
MOUND CITY WARRIOR, Brown Shoe Co., St. Louis, Mo
MOUND CITY, WORK WELT, Brown Shoe Co., St. Louis,
MOUNTAIN, Hamilton-Brown Shoe Co., St. Louis, Mo.
MOUNTAIN, Friedman-Shelby Shoe Co., St. Louis, Mo.
MURRAY SHOE CO., Murray Shoe Co., Lynn, Mass.
MUSIC, Roberts, Johnson & Rand Shoe Co., St. Louis, Mo.
MYRA, Friedman-Shelby Shoe Co., St. Louis, Mo.
N. & N., Green-Wheeler Shoe Co., Ft. Dodge, Iowa.
"MUDGE", James Clark Leather Co., St. Louis, Mo.
NATION'S WHITE HOUSE, Brown Shoe Co., St. Louis, Mo
NATIVE DAUGHTER, G. M. Kutz Shoe Co., San Francisco, California.
NEMO, Friedman-Shelby Shoe Co., St. Louis, Mo.
NESMITH, Nesmith Shoe Co., Brockton, Mass.
NETTLETON, A. E. Nettleton Co., Syracuse, N. Y.

NEW CENTURY, A. H. Berry Shoe Co., Portland, Maine.
NEW IDEA, Dittmann Shoe Co., St. Louis, Mo.
NEWMARKET, Endicott, Johnson & Co., Endicott, N. Y.
NEW MEXICO. Peters Shoe Co., St. Louis, Mo.
NEW ROYAL, The Cady-Ivison Shoe Co., Cleveland, Ohio.
NEW THOUGHT SHOE, The Plant-Butler Co., Cincinnati, Ohio.
NEW WINNER, Brown Shoe Co., St. Louis, Mo.
NIAGARA, Peters Shoe Co., St. Louis, Mo.
NIAGARA, H. D. Raff & Co., Chicago, Ill.
NICKLE PLATE, The Cady-Ivison Shoe Co., Cleveland, Ohio.
NINE O'CLOCK, Dittmann Shoe Co., St. Louis, Mo.
NIT TRIP, Dugan & Hudson Co., Rochester, N. Y.
NOKORN SHOE, O'Sullivan Bros. Co., Lowell, Mass.
NOR-CUSH, Northern Shoe Co., Duluth, Minn.
NORTHLAND, Roberts, Johnson & Rand Shoe Co., St. Louis, Mo.
NORTHMORE, Ellet-Kendall Shoe Co., Kansas City, Mo.
NORTH POLE, Roberts, Johnson & Rand Shoe Co., St. Louis, Mo.
NO TAX, Geddes-Brown Shoe Co., Indianapolis, Ind.
NOW THEN, Peters Shoe Co., St. Louis, Mo.
NOXALL, Dittmann Shoe Co., St. Louis, Mo.
NU BABY, Nu Baby Shoe Co., Lynn, Mass.
NUTCRACKER, Roberts, Johnson & Rand Shoe Co., St. Louis, Mo.
O. K., Dittmann Shoe Co., St. Louis, Mo.
O. K. Brown Shoe Co., St. Louis, Mo.
O. S. U., The Henry C. Werner Co., Columbus, Ohio.
O-U-SHU, Peters Shoe Co., St. Louis, Mo.
O'DONNELL, O'Donnell Shoe Co., St. Paul, Minn.
OHIO, Dittmann Shoe Co., St. Louis, Mo.
OHIO, Brown Shoe Co., St. Louis, Mo.
OLD GLORY, Geo. H. Snow Co., Brockton, Mass.
OLD HOMESTEAD, Rice & Hutchins, Boston, Mass.
OLD HONESTY, Filsinger-Boette Shoe Co., St. Louis, Mo .
OLD HICKORY, Filsinger-Boette Shoe Co., St. Louis, Mo.
OLD ORIGINAL, Elk Skin Moccasin Mfg. Co., Ypsilanti, Mich.
OLYMPIA, Friedman-Shelby Shoe Co., St. Louis, Mo.
"ONCE ALFEAR," J. K. Orr Shoe Co., Atlanta, Ga.
OPENER, Endicott, Johnson & Co., Endicott, N. Y.
OREGON, Hamilton-Brown Shoe Co., St. Louis, Mo.
OREGON, Friedman-Shelby Shoe Co., St. Louis, Mo.
ORIGINAL CHIPPEWA, Chippewa Shoe Mfg. Co., Chippewa. Falls, Wis.
ORIGINAL 493, Dittmann Shoe Co., St. Louis, Mo.
ORIOLE, Brown Shoe Co., St. Louis, Mo.

OSAGE, Peters Shoe Co., St. Louis, Mo.
OUR ALICE, Bode-Larson Shoe Co., Keokuk, Iowa.
OUR CONGRESSMAN, Ellet-Kendall Shoe Co., Kansas City, Mo.
OUR FAMILY, Roberts, Johnson & Rand Shoe Co., St. Louis, Mo.
OUR OWN CUSHION SHOE, Utz & Dunn Co., Rochester, N. Y.
OUR PARAGON, Bode-Larson Shoe Co., Keokuk, Ia.
OUR SPECIAL, The Cady-Ivison Shoe Co., Cleveland, Ohio
OUTING, Rice & Hutchins, Boston, Mass.
OUTING, Marion Shoe Co., Marion, Ind.
OUTING, Barker, Brown & Company, Huntington, Ind.
OUTING, Peters Shoe Co., St. Louis, Mo.
OVERLAND, Bode-Larson Shoe, Keokuk, Iowa.
OVERTON PARK, Carruthers-Jones Shoe Co., St. Louis, Mo.
OWL LINE, Noyes-Norman Shoe Co., St. Joseph, Mo.
OWNLEE, S. Bachardt & Co., New York, N. Y.
OWN MAKE, Hamilton-Brown Shoe Co., St. Louis, Mo.
OX CALF, Brown Shoe Co., St. Louis, Mo.
OX CALF MOUND CITY, Brown Shoe Co., St. Louis, Mo.
OZARK, Friedman-Shelby Shoe Co., St. Louis, Mo.
PACEMAKER, Roberts, Johnson & Rand Shoe Co., St. Louis, Mo.
PACE SETTER, Peters Shoe Co., St. Louis, Mo.
PACIFIC SPECIAL, Friedman-Shelby Shoe Co., St. Louis, Mo.
PACKARD SHOE, M. A. Packard Co., Brockton, Mass.
PAINLESS TURNS, Xenia Shoe Mfg. Co., Xenia, Ohio.
PAIR THAT WEAR, The Cady-Ivison Shoe Co., Cleveland, Ohio.
PALMA, Friedman-Shelby Shoe Co., St. Louis, Mo.
PAN HANDE, Friedman-Shelby Shoe Co., St. Louis, Mo.
PANNANT, Peters Shoe Co., St. Louis, Mo.
PAPILLION, C. P. Ford & Co., Rochester, N. Y.
PAPOOSE, Brown Shoe Co., St. Louis, Mo.
PARAGON, Faunce & Spinney, Lynn, Mass.
PARAGON, Bentley & Olmsted Co., Des Moines, Iowa.
PAR EXCELLENCE, Peters Shoe Co., St. Louis, Mo.
PARIS, Dougherty-Fithian Shoe Co., Portland, Oregon.
PAR VALUE, Brown Shoe Co., St. Louis, Mo.
PATRICIAN, Faunce & Spinney Lynn, Mass.
PATRIOT, Roberts, Johnson & Rand Shoe Co., St. Louis, Mo.
PAYNE'S WEST VIRGINIA SHOE, Payne Shoe Co., Charleston, W. Va.
PEACH, Roberts, Johnson & Rand Shoe Co., St. Louis, Mo.
PEACH, Dittmann Shoe Co., St. Louis, Mo.
PECAS, Friedman-Shelby Shoe Co., St. Louis, Mo.
PEDESTRIAN, Barton Bros., Kansas City, Mo.

PEDINEAT, Blum Shoe Mfg. Co., Dansville, N. Y.
PEDOMIK, Rice & Hutchins, Boston, Mass.
PEEK-A-BOO, Brown Shoe Co., St. Louis, Mo.
PEERLEE, Daniel Green Felt Shoe Co., Dolgeville, N. Y.
PEERLESS, Harrisburg Shoe Mfg. Co., Harrisburg, Pa.
PEERLESS, S. Bachardt & Co., New York, N. Y.
PEMBERTON, Friedman-Shelby Shoe Co., St. Louis, Mo.
PERFECT MODEL, Lewis A .Crossett, Inc., North Abington,
    Mass.
PERFECTION, Noyes Norman Shoe Co., St. Joseph, Mo.
PET, Peters Shoe Co., St. Louis, Mo.
PET, Brown Shoe Co., St. Louis, Mo.
PIAZZA, Brown Shoe Co., St. Louis, Mo.
PICKWICK, The Apex Shoe Factory, New Orleans, La.
PICNIC, Hamilton-Brown Shoe Co., St. Louis, Mo.
PILGRIM, Roberts, Johnson & Rand Shoe Co., St. Louis, Mo.
PINE KNOT, Friedman-Shelby Shoe Co., St. Louis, Mo.
PINE KNOT, Hamilton-Brown Shoe Co., St. Louis, Mo.
PINGREE COMFORT, The Pingree Company, Detroit, Mich.
PINGREE-MADE, The Pingree Company, Detroit, Mich.
PINGREE-MADE CUSHION, Pingree Company, The, De-
    troit, Mich.
PINK, Roberts, Johnson & Rand Shoe Co., St. Louis, Mo.
PIONEER, Excelsior Shoe & Slipper Co., Cedarburg, Wis.
PIONEER, Herold Bertsck Shoe Co., Grand Rapids, Mich.
PIVOT, Friedman-Shelby Shoe Co., St. Louis, Mo.
PHONOGRAPH, Lynchburg Shoe Co., Inc., Lynchburg, Va
PLAIN TRUTH, Peters Shoe Co., St. Louis, Mo.
PLA-MATE, Williams, Hoyt & Co., Rochester, N. Y.
PLAY DAY, Peters Shoe Co., St. Louis, Mo.
PLAY DAY, Brown Shoe Co., St. Louis, Mo.
PLAYFAIR, Foot, Schulze & Co., St. Paul, Minn.
PLAYMATE, Hirth-Krause Company, Grand Rapids, Mich.
PLEASURE, Friedman-Shelby Shoe Co., St. Louis, Mo.
PLUCK, Roberts, Johnson & Rand Shoe Co., St. Louis, Mo.
POCO, Friedman-Shelby Shoe Co., St. Louis, Mo.
POLAR STAR, Roberts, Johnson & Rand Shoe Co., St. Louis,
    Mo.
PONTIAC'S WEAR WELL, Pontiac Shoe Mfg. Co., Pontiac,
    Ill.
POPULARIS, Harrisburg Shoe Mfg. Co., Harrisburg, Pa.
PORTSMOUTH, The Portsmouth Shoe Co., Portsmouth, O.
PREDMONT, Friedman-Shelby Shoe Co., St. Louis, Mo.
PREMIER, J. E. Tilt Shoe Co.. Chicago, Ill.
PRESIDENT WHITE HOUSE, Brown Shoe Co., St. Louis,
    Mo.
PRIDE, Friedman-Shelby Shoe Co., St. Louis, Mo.
PRINCE KARL, Geddes-Brown Shoe Co., Indianapolis, Ind.

PRINCESS, Friedman-Shelby Shoe Co., St. Louis, Mo.
PRINCESS, Nathan D. Dodge Shoe Co., Newburyport, Mass.
PRINCESS, Brown Shoe Co., St. Louis, Mo.
PRINCESS, Peters Shoe Co., St. Louis, Mo.
PRINCESS, H. Lindenberg, New Orleans, La.
PRINCESS HENRY, The Cady-Ivison Shoe Co., Cleveland, Ohio.
PRINCESS LOUISE, A. H. Berry Shoe Co., Portland, Maine.
PRINCETON, Dougherty-Fithian Shoe Co., Portland, Oregon.
PRISCILLA, The Henry C. Werner Co., Columbus, Ohio.
PROMENADE, Peters Shoe Co., St. Louis, Mo.
PROSPECTOR, Roberts, Johnson & Rand Shoe Co., St. Louis Mo
PROTECTOR, Peters Shoe Co., St. Louis, Mo.
PURE, Friedman-Shelby Shoe Co., St. Louis, Mo.
PURPLE COW, Filsinger-Boette Shoe Co., St. Louis, Mo.
PUTNAM, Geddes-Brown Shoe Co., Indianapolis, Ind.
PYRAMID, Friedman-Shelby Shoe Co., St. Louis, Mo.
QUAKER, Roberts, Johnson & Rand Shoe Co., St. Louis, Mo.
QUAKER, Hamilton-Brown Shoe Co., St. Louis, Mo.
QUALITY, Carruthers-Jones Shoe Co., St. Louis, Mo.
QUANAH, Friedman-Shelby Shoe Co., St. Louis, Mo.
QUEEN B., Brown Shoe Co., St. Louis, Mo.
"QUEEN BESS", J. K. Orr Shoe Co., Atlanta, Ga.
QUEEN OF THE WEST, Ellet-Kendall Shoe Co., Kansas City, Mo.
QUEEN QUALITY, Thomas G. Plant Co., Boston, Mass.
QUEEN ROSALIND, Jos. Rosenheim Shoe Co., Savannah, Ga.
QUEEN TASTE, Jos. Rosenheim Shoe Co., Savannah, Ga.
QUICKSTEP, Peters Shoe Co., St. Louis, Mo.
QUINCY, Dittmann Shoe Co., St. Louis, Mo.
R. D. G., Reynolds, Drake & Gabell Co., N. Easton, Mass.
R-E-Z, Sharood Shoe Corporation, St. Paul Minn.
R. J. & R., Roberts, Johnson & Rand Shoe Co., St. Louis, Mo.
R. & G. SHOE, The Richard-Gregory Shoe Co., Lynn, Mass.
R. & H., Rice & Hutchins, Boston, Mass.
R. & H. Special, Rice & Hutchins, Boston, Mass.
RACINE, Racine Shoe Co., Racine, Wis.
RADCLIFFE, The Radcliffe Shoe Co., Boston, Mass.
RAILROADER, Endicott, Johnson & Co., Endicott, N. Y.
RAINIER, Northwestern Shoe Co., Seattle, Wash.
RALSTON, Churchill & Alden Co., also Ralston Health Shoe-makers, Campello, Mass.
RAMBLER, Hamilton-Brown Shoe Co., St. Louis, Mo.
RANGER, Hamilton-Brown Shoe Co., St. Louis, Mo.
RAPID TRANSIT, The Cady-Ivison Shoe Co., Cleveland, Ohio.
REAL EASE, Ellet-Kendall Shoe Co., Kansas City, Mo.

REAL REST, Brown Shoe Co., St. Louis, Mo.
"REAL SHOE MAKERS", Forbush Shoe Co., North Grafton, Mass.
REAPER, Hamilton-Brown Shoe Co., St. Louis, Mo.
RECESS SCHOOL SHOES, Ellet-Kendall Shoe Co., St. Louis, Mo.
RED BIRD, Peters Shoe Co., St. Louis, Mo.
RED CROSS, Krohn-Fecheimer Co., Cincinnati, Ohio.
RED GOOSE, Friedman-Shelby Shoe Co., St. Louis, Mo.
RED LION, Barker, Brown & Co., Huntington, Ind.
RED PEPPER, Roberts, Johnson & Rand Shoe Co., St. Louis, Mo.
RED RIVER, Dittmann Shoe Co., St. Louis, Mo.
RED RIVER, Roberts, Johnson & Rand Shoe Co., St. Louis, Mo.
REED'S, CUSHION SHOE, DR. A., John Ebberts Shoe Co., Buffalo, N. Y.
REED SHOE, THE, H. B. Reed & Co., Manchester, N. H.
REGAL, Regal Shoe Co., Whitman, Mass.
REGENT, The Cady-Ivison Shoe Co., Cleveland, Ohio.
REGENT, Helmers-Bettman, Cincinnati, Ohio.
REGENT, Friedman-Shelby Shoe Co., St. Louis, Mo.
REGENT RULER, Jos. Rosenheim Shoe Co., Savannah, Ga.
REGINA, Buchanan-Lawrence Co., Joliet, Ill.
REGULATOR, Hamilton-Brown Shoe Co., St. Louis, Mo.
REIGNING QUEEN, Arnold-Henegar-Doyle Co., Knoxville, Tenn.
REILY SHOE, Crowder-Cooper Shoe Co., Indianapolis, Ind.
REILY'S BABIES, Crowder-Cooper Shoe Co., Indianapolis, Ind.
RE-LAX SHOE, Hayward Bros. Shoe Co., Omaha, Neb.
RELIANCE, Dittmann Shoe Co., St. Louis, Mo.
RELIANCE, H. Pretzfelder & Co., Baltimore, Md.
RELIANCE, A. Priesmeyer Shoe Co., Jefferson City, Mo.
RELIO, A. S. Kreider Shoe Co., Annville, Pa.
RELIO, The Kreider, Schwarz & Sallenbach Shoe Co., St. Louis, Mo.
REO, Friedman-Shelby Shoe Co., St. Louis, Mo.
REPEATER, Brown Shoe Co., St. Louis, Mo.
REPUBLIC SCHOOL SHOES, Bentley & Olmsted Co., Des Moines, Iowa.
RESISTALL, Dittmann Shoe Co., St. Louis, Mo.
RESTER, Milford Shoe Co., Milford, Mass.
RESTFUL, Hirth-Krause Co., Grand Rapids, Mich.
REVELATION, Banner Shoe Company, St. Louis, Mo.
REVERE, Nu Baby Shoe Co., Lynn, Mass.
REX, Dittmann Shoe Co., St. Louis, Mo.
"REX", American Specialty Shoe Co., Milwaukee, Wis.
REX CALF, Rice & Hutchins, Boston, Mass.

36

REYNOLDS, Reynolds, Drake & Gabell Co., N. Easton, Mass.
RICHMOND, Brown Shoe Co., St. Louis, Mo.
RIDGEWOOD, Crowder-Cooper Shoe Co., Indianapolis, Ind.
RIGHT DRESS, Lynchburg Shoe Co., Inc., Lynchburg, Va.
RIGHT ROYAL, Jos. Rosenheim Shoe Co., Savannah, Ga.
RIGHT ROYAL, The Henry C. Werner Co., Columbus, Ohio.
RIGHTWAY, The Scheiffele Shoe Mfg. Co., Cincinnati, Ohio.
RIGI, Geddes-Brown Shoe Co., Indianapolis, Ind.
RIGOR KID, Green-Wheeler Shoe Co., Ft. Dodge, Iowa.
RINGER, Peters Shoe Co., St. Louis, Mo.
RINGWOOD, Friedman-Shelby Shoe Co., St. Louis, Mo.
RISING SUN, The Cady-Ivison Shoe Co., Cleveland, Ohio.
RITA, Friedman-Shelby Shoe Co., St. Louis, Mo.
RIVAL, The Cady-Ivison Shoe Co., Cleveland, Ohio.
RIVAL, Noyes-Norman Shoe Co., St. Joseph, Mo.
RIVELATION, Howard, Briggs & Pray Co., Auburn, Maine.
ROBERTSON, Brown Shoe Co., St. Louis, Mo.
ROCHESTER MADE, H. D. Raff & Co., Chicago, Ill.
ROCK, Brown Shoe Co., St. Louis, Mo.
ROCKBOTTOM, O'Sullivan Bros. Co., Lowell, Mass.
ROCK CRUSHER, Barker, Brown & Co., Huntington, Ind.
ROCK HILL, Dougherty-Fithian Shoe Co., Portland, Oregon.
ROCKY MOUNTAIN, Foot, Schulze & Co., St. Paul, Minn.
ROCK OF GIBRALTER, Hamilton- Brown Shoe Co., St.
   Louis, Mo.
ROPER, Roberts, Johnson & Rand Shoe Co., St. Louis, Mo.
ROSE BUD, Crowder-Cooper Shoe Co., Indianapolis, Ind.
ROSEBUD, Friedman-Shelby Shoe Co., St. Louis, Mo.
ROSE CITY, Dougherty-Fithian Shoe Co., Portland, Oregon.
ROSE OF K. C., Barton Bros., Kansas Ctiy, Mo.
ROSEMARY, Crowder-Cooper Shoe Co., Indianapolis, Ind.
ROSES, Roberts, Johnson & Rand Shoe Co., St. Louis, Mo.
ROTARY, The Washington Shoe Mfg. Co., Seattle, Wash.
ROYAL, Sawyer Boot & Shoe Co., Bangor, Maine.
ROYAL, Huiskamp Bros. Co., Keokuk, Iowa.
ROYAL, The Rich Shoe Co., Milwaukee, Wis.
ROYAL BLUE, Selz-Schwab Shoe Co., Chicago, Ill.
ROYAL CALF, Dittmann Shoe Co., St. Louis, Mo.
ROYAL CREST, The Cady-Ivison Shoe Co., Cleveland, Ohio.
ROMAN, Brown Shoe Co., St. Louis, Mo.
ROMEOS SLIPPERS, H. Lindenberg, New Orleans, La.
ROMPER, Utz & Dunn Co., Rochester, N. Y.
ROMPER, Brown Shoe Co., St. Louis, Mo.
RONGE REX, Hirth-Krause Co., Grand Rapids, Mich.
ROUGH & READY, Friedman-Shelby Shoe Co., St. Louis, Mo.
ROUGH RIDER, Excelsior Shoe & Slipper Co., Cedarburg,
   Wis.
ROUGH RIDER, Brown Shoe Co., St. Louis, Mo.

ROUND-UP, Brown Shoe Co., St. Louis, Mo.
RUNABOUT, J. P. Hartray Shoe Co., Chicago, Ill.
RUNAWAY GIRL, Lynchburg Shoe Co., Inc., Lynchburg, Va.
RUGGED, Endicott, Johnson & Co., Endicott, N. Y.
RUGGED, Friedman-Shelby Shoe Co., St. Louis, Mo.
RUTH, Hirth-Krause Co., Grand Rapids, Michigan.
S. & M. Slater & Morrill, Inc., South Braintree, Mass.
S. & M. JUNIORS, Slater & Morrill, Inc., ·South Braintree, Mass.
S. & S., The Kreider, Schwarz & Sallenbach Shoe Co., St. Louis, Mo.
ST. CECILIA, Utz & ·Dunn Co., Rochester, N. Y.
ST. CRISPIN, Filsinger-Boette Shoe Co., St. Louis, Mo.
ST. LOUIS MAID, Wertheimer-Swarts Shoe Co., St. Louis, Mo.
ST. REGIS, H. Pretzfelder & Co., Baltimore, Md.
SALAMANDER, C. W. Johnson, Natick, Mass.
SALLY WALKER, The Scheiffele Shoe Mfg. Co., Cincinnati, Ohio.
SAMAROFF, Nathan D. Dodge Shoe Co., Newburyport, Mass.
SANDOW, Filsinger-Boette Shoe Co., St. Louis, Mo.
SANITARY, Friedman-Shelby Shoe Co., St. Louis, Mo.
SANITOS, Sharood Shoe Corporation, St. Paul, Minn.
SANTA CLARA, Roberts, Johnson & Rand Shoe Co., St. Louis, Mo.
SANTA FE, Dittmann Shoe Co., St. Louis, Mo.
SANTA ROSA, Roberts, Johnson & Rand Shoe Co., St. Louis, Mo.
SATIN, Roberts, Johnson & Rand Shoe Co., St. Louis, Mo.
SATISFIER, Jos. Rosenheim Shoe Co., Savannah, Ga.
SAXONIA, The Henry C. Werner Co., Columbus, Ohio.
SCHOOL DAYS, Xenia Shoe Mfg. Co., Xenia, Ohio.
SCHOOL-GIRL, Endicott, Johnson & Co., Endicott, N. Y.
SCHOOLMATE, Guthmann, Carpenter & Telling, Chicago, Ill.
SCHOOL-MATE, Williams, Hoyt & Co., Rochester, N. Y.
SCIENTIFIC, S. Bachardt & Co., New York, N. Y.
SCRANTON, Endicott, Johnson & Co., Endicott, N. Y.
SEALED SOLE, The Pingree Co., Detroit, Mich.
SECURITY, Endicott, Johnson & Co., Endicott, N. Y.
SECURITY, Hamilton-Brown Shoe Co., St. Louis, Mo.
SECURITY SCUFFER, Hamilton-Brown Shoe Co., St. Louis, Mo.
SELBY'S "FLEX-O", The Shelby Shoe Co., Portsmouth, O.
SELBY, Selby Shoe Co., Portsmouth, Ohio.
SELKIRK, F. P. Kirkendall & Co., Omaha, Neb.
SENATE, Friedman-Shelby Shoe Co., St. Louis, Mo.
SENSATION, Roberts, Johnson & Rand Shoe Co., St. Louis, Mo.

SENATOR, THE. Ellet-Kendall Shoe Co., St. Louis, Mo.
SERVICE, Roberts, Johnson & Rand Shoe Co., St. Louis, Mo.
SERVICE, Endicott, Johnson & Co., Endicott, N. Y.
SIGNET, Rice & Hutchins, Boston, Mass.
SIR KNIGHT, Wertheimer-Swarts Shoe Co., St. Louis, Mo.
SITTING BULL, Elk Skin Moccasin Mfg. Co., Ypsilanti, Mich.
SHAMROCK, Arnold-Henegar-Doyle Co., Knoxville, Tenn.
SHE, Carruthers-Jones Shoe Co., St. Louis, Mo.
SHOD FOR A YEAR. Hamilton-Brown Shoe Co., St. Louis,
    Mo.
SHOD-WELL, The Jos. P. Dunn Shoe & Leather Co., Denver,
    Colo
SHOP MADE, Roberts, Johnson & Rand Shoe Co., St. Louis,
    Mo.
SHOP MADE, Hamburger Bros. Shoe Co., St. Louis, Mo.
SHU-EZE, Hamilton-Brown Shoe Co., St. Louis, Mo.
SHUMATES, Whittinghill-Harlow Shoe Co., St. Joseph, Mo.
SHU-RITE, The Jos. P. Dunn Shoe & Leather Co., Denver,
    Colo.
SKAPPER, Williams Hoyt & Co., Rochester, N. Y.
SKI HI, The Plaut Butler Co., Cincinnati, Ohio.
SKREENMER, Fred F. Field Co., Brockton, Mass.
"SKREEMER", Payne Shoe Co., Charleston, W. Va.
SKUFFERS, Engel-Cone Shoe Co., Boston, Mass.
"SNOW", Geo. H. Snow Co., Brockton, Mass.
SOCIETY, Roberts, Johnson & Rand Shoe Co., St. Louis, Mo.
SOCIETY KING, Lynchburg Shoe Co., Inc., Lynchburg, Va.
SOCIETY STAR, Roberts, Johnson & Rand Shoe Co., St.
    Louis, Mo.
SO ESY, Brown Shoe Co., St. Louis, Mo.
SOFT & GOOD, Roberts, Johnson & Rand Shoe Co., St. Louis,
    Mo.
SOFT & STRONG, Huiskamp Bros. Co., Keokuk, Iowa.
SOLASTIC, Wertheimer-Swarts Shoe Co., St. Louis, Mo.
SOLE EASE, The Bering Shoe Co., Cincinnati, Ohio.
SOLID, The Cady-Ivison Shoe Co., Cleveland, Ohio.
SOLID ROCK, Hinkle Shoe Co., Evansville, Ind.
SOLO, Brown Shoe Co., St. Louis, Mo.
SO-NETE, Whittinghill-Harlow Shoe Co., St. Joseph, Mo.
SOROSIS, A. E. Little & Co., Lynn, Mass.
SOUTHERN GIRL, Craddock Terry Co., Lynchburg, Va.
SOUTHERN QUEEN, Brown Shoe Co., St. Louis, Mo.
SOUTHLAND BELL, Craddock-Terry Co., Lynchburg, Va.
SOVERIGN QUALITY, The Cady-Ivison Shoe Co., Cleveland,
    Ohio.
STANDPAT, Wertheimer-Swarts Shoe Co., St. Louis, Mo.
STAR, Star Baby Shoe Co., Minneapolis, Minn.
STAR, Harriburg Shoe Mfg. Co., Harrisburg, Pa.

STAR CALF, Roberts. Johnson & Rand Shoe Co., St. Louis, Mo.
STAR CALF, Roberts, Johnson & Rand Shoe Co., St. Louis, Mo.
STAR-FIVE-STAR, Brown Shoe Co., St. Louis, Mo.
STAR-OF-THE-WEST, E. B. Pickenbrock & Sons, Dubuque. Iowa.
STAR POPULARIS, Harrisburg Shoe Mfg. Co., Harrisburg, Pa.
STARTER, Peters Shoe Co., St. Louis, Mo.
STARTRIGHT, Geo. Baker & Sons, Inc., Brooklyn, N. Y.
STATESMAN, Friedman-Shelby Shoe Co., St. Louis, Mo.
STEADFAST. Smith-Briscoe Shoe Co., Inc., Lynchburg, Va.
STEEL SHOD, Brennand & White, Brooklyn, N. Y.
STERLING, Banner Shoe Company, St. Louis, Mo.
STERLING, Endicott, Johnson & Co., Endicott, N. Y.
STERLING, Brown Shoe Co., St. Louis, Mo.
STERLING QUALITY, Noyes Norman Shoe Co., St. Joseph, Mo.
STETSON. Stetson Shoe Co., South Weymouth, Mass.
STEWART'S BEST, The Cady-Ivison Shoe Co., Cleveland, Ohio.
STONE CRUSHER, The Cady-Ivison Shoe Co., Cleveland, Ohio.
STONEWALL, Filsinger-Boette Shoe Co., St. Louis, Mo.
STORM, Peters Shoe Co., St. Louis, Mo.
STRENUOUS, Brown Shoe Co., St. Louis, Mo.
STRONG BOY, Geddes-Brown Shoe Co., Indianapolis, Ind.
STRONG-HEART, Endicott, Johnson & Co., Endicott, N. Y.
STRONGHOLD, Roberts, Johnson & Rand Shoe Co., St. Louis, Mo.
STRONGHOLD, Dittmann Shoe Co., St. Louis, Mo.
STRONGER-THAN-THE-LAW, Roberts, Johnson & Rand Shoe Co., St. Louis, Mo.
STROOTMAN, John Strootman Shoe Co., Buffalo, N. Y.
STUBPROOF, Endicott, Johnson & Co., Endicott, N. Y.
STUMP OF THE WORLD, Brown Shoe Co., St. Louis, Mo.
STUMP OF THE WORLD. Brown Shoe Co., St. Louis, Mo.
STURDY, Dittmann Shoe Co., St. Louis, Mo.
STURDY, Endicott, Johnson & Co., Endicott, N. Y.
STYLE, Roberts, Johnson & Rand Shoe Co., St. Louis, Mo.
STYLE, Dougherty-Fithian Shoe Co., Portland, Oregon.
SUCCESS, Brown Shoe Co., St. Louis, Mo.
SUCCESS, Friedman-Shelby Shoe Co., St. Louis, Mo.
SUCCESS, Hamilton-Brown Shoe Co., St. Louis, Mo.
SULTANA, Buchanan-Lawrence Co., Joliet, Ill.
SULTANA, Daniel Green Felt Shoe Co., Dolgeville, N. Y.
SUMMER, Friedman-Shelby Shoe Co., St. Louis, Mo.

SUMMERSET, Craddock Terry Co., Lynchburg, Va.
SUMMIT, Churchill & Alden Co., also Ralston Health Shoe-
makers, Campello, Mass.
SUM-SHU, Peters Shoe Co., St. Louis, Mo.
SUNBEAM, Roberts, Johnson & Rand Shoe Co., St. Louis, Mo.
SUNBEAM, Brown Shoe Co., St. Louis, Mo.
SUNFLOWER, Noyes-Norman Shoe Co., St. Joseph, Mo.
SUNSET, Calm, Nickelsburg & Co., San Francisco, Cal.
SUNSET, Hamilton-Brown Shoe Co., St. Louis, Mo.
SUNSET, H. Pretzfelder & Co., Baltimore, Md.
SUNSHINE, Endicott, Johnson & Co., Endicott, N. Y.
SUNNY JIM, Jos. Rosenheim Shoe Co., Savannah, Ga.
SUNNY SOUTH, Friedman-Shelby Shoe Co., St. Louis, Mo.
SUSETTE, Daniel Green Felt Shoe Co., Dolgeville, N. Y.
SUPERBA, The Henry C. Werner Co., Columbus, Ohio.
SUPERIOR, The Cady-Ivison Shoe Co., Cleveland, Ohio.
SUPREMA, Harrisburg Shoe Mfg. Co., Harrisburg, Pa.
SURE POP, Friedman-Shelby Shoe Co., St. Louis, Mo.
SURPASS, A. Priesmeyer Shoe Co., Jefferson City, Mo.
SURVEYER, Friedman-Shelby Shoe Co., St. Louis, Mo.
SURVEYOR, Hamilton-Brown Shoe Co., St. Louis, Mo.
SPARKLER, Peters Shoe Co., St. Louis, Mo.
SPARTAN, Geo. F. Daniels & Co., Boston, Mass.
"SPEAR BRAND," Spear Bros. Co., Baltimore, Md.
SPECIAL MERIT, F. Mayer Boot & Shoe Co., Milwaukee,
Wis.
SPECIAL SERVICE, The Starner-Copeland Co., Columbus,
Ohio.
"SPENCERIA SPECIAL", E. Jones & Co., Inc., Spencer,
Mass.
SPINDLE TOP, Friedman-Shelby Shoe Co., St. Louis, Mo.
SPINDLE TOP, Hamilton-Brown Shoe Co., St. Louis, Mo.
SPLIT, Endicott, Johnson & Co., Endicott, N. Y.
SPORTSMAN, Dittmann Shoe Co., St. Louis, Mo.
SPORTSMAN, Hamilton-Brown Shoe Co., St. Louis, Mo.
SPRINTER, Wood & Johnson Co., Rochester, N. Y.
SWEET ECHO, Jos. Rosenheim Shoe Co., Savannah, Ga.
SWEETHEART, Peters Shoe Co., St. Louis, Mo.
SWEETHEART, Dittmann Shoe Co., St. Louis, Mo.
SWELL, Friedman-Shelby Shoe Co., St. Louis, Mo.
TAILOR MADE, The Washington Shoe Mfg. Co., Seattle,
Wash.
TAILOR MADE, Daniel Green Felt Shoe Co., Dolgeville,
N. Y.
TAMARACK, Northern Shoe Co., Duluth, Minn.
TAN ELK, Friedman-Shelby Shoe Co., St. Louis, Mo.
TAPPANS, Tappan Shoe Mfg. Co., Coldwater, Mich.
TEDDY BEAR, Peters Shoe Co., St. Louis, Mo.

"TENDERFOOT", The Starner-Copeland Co., Columbus, O.
TENNIS, Peters Shoe Co., St. Louis, Mo.
TERROR, Hamilton-Brown Shoe Co., St. Louis, Mo.
TERRY SPECIAL, Craddock Terry Co., Lynchburg, Va.
TESS & TEDD, Roberts, Johnson & Rand Shoe Co., St. Louis, Mo.
THOMPSON SHOE, Thompson Bros., Campello, Mass.
THUNDERBOLT, Jos. Rosenheim Shoe Co., Savannah, Ga.
TIFFANY, Foss, Packand & Co., Boston, Mass.
TIGER, P. W. Minor & Son, Batavia, N. Y.
TINY TOT'S, Hyman Bros., Rochester, N. Y.
TIP TOP, Peters Shoe Co., St. Louis, Mo.
TOBASCO, Hamilton-Brown Shoe Co., St. Louis, Mo.
TOIL, Hamilton-Brown Shoe Co., St. Louis, Mo.
TOILER, Brown Shoe Co., St. Louis, Mo.
TOM BOY, Burrow, Jones & Dyer Shoe Co., St. Louis, Mo.
TOPEKA, Friedman-Shelby Shoe Co., St. Louis, Mo.
TOPSY, Friedman-Shelby Shoe Co., St. Louis, Mo.
TORNADO, Peters Shoe Co., St. Louis, Mo.
TORPEDO, Hamilton-Brown Shoe Co., St. Louis, Mo.
TOSCA, Daniel Green Felt Shoe Co., Dolgeville, N. Y.
TOT, Peters Shoe Co., St. Louis, Mo.
TOURIST, H. Pretzfelder & Co., Baltimore, Md.
TOWER, Friedman-Shelby Shoe Co., St. Louis, Mo.
TOWN TALK, A. Priesmeyer Shoe Co., Jefferson, City, Mo.
TRIANGLE, Friedman-Shelby Shoe Co., St. Louis, Mo.
TRIANGLE, Hayward Bros. Shoe Co., Omaha, Neb.
TRIFLE, Friedman-Shelby Shoe Co., St. Louis, Mo.
TRINIDAD, Brown Shoe Co., St. Louis, Mo.
TRINITY, Buchanan-Lawrence Co., Joliet, Ill.
TRIPLE WEAR, Marion Shoe Co., Marion, Ind.
TRIUMPH, The Cady-Ivison Shoe Co., Cleveland, Ohio.
TROPHY, Brown Shoe Co., St. Louis, Mo.
TRUEASE, Guthmann, Carpenter & Telling, Chicago, Ill.
TRUFORM, O'Sullivan Bros. Co., Lowell, Mass.
TRYONDA, A. Priesmeyer Shoe Co., Jefferson City, Mo.
TREADWELL, F. Mayer Boot & Shoe Co., Milwaukee, Wis.
TREAD-WELL, The Washington Shoe Mfg. Co., Seattle, Wash.
TREADEASY, P. W. Minor & Son, Batavia, N. Y.
TREMONT, Burrow, Jones & Dyer Shoe Co., St. Louis, Mo.
TRADE, Hamilton- Brown Shoe Co., St. Louis, Mo.
TRADE LINE, Endicott, Johnson & Co., Endicott, N. Y.
TRAVELER, Friedman-Shelby Shoe Co., St. Louis, Mo.
TRAVELERS, Nolan-Earl Shoe Co., San Francisco, Cal.
TUFFER, Peters Shoe Co., St. Louis, Mo.
TUF-FIR, The Washington Shoe Mfg. Co., Seattle, Wash.
TUFF WEAR, Dittmann Shoe Co., St. Louis, Mo.

TULIP, Peters Shoe Co., St. Louis, Mo.
TURN-OR-WELT, A. H. Berry Shoe Co., Portland, Maine.
TUTTLE, Morse & Rogers, New York, N. Y.
U-CANT- BUST-M, The Kreider, Schwarz & Sallenbach Shoe
    Co., St. Louis, Mo.
UMPIRE, Roberts, Johnson & Rand Shoe Co., St. Louis, Mo.
UNAPPROACHABLE, The Cady-Ivison Shoe Co., Cleveland,
    Ohio.
UNCLE JOSH, Noyes Norman Shoe Co., St. Joseph, Mo.
UNCLE SAM, Filsinger-Boette Shoe Co., St. Louis, Mo.
U-NEED-AN-ELK, The Cady-Ivison Shoe Co., Cleveland, O.
UNIQUE, Brown Shoe Co., St. Louis, Mo.
UNIQUE SPECIAL, Brown Shoe Co., St. Louis, Mo.
UNIVERSAL, A. Priesmeyer Shoe Co., Jefferson City, Mo.
UNIVERSITY, Guthmann, Carpenter & Telling, Chicago, Ill.
UNIVERSITY, Banner Shoe Company, St. Louis, Mo.
U. S., Friedman-Shelby Shoe Co., St. Louis, Mo.
U. S. A., A. Priesmeyer Shoe Co., Jefferson City, Mo
USONA, Brown Shoe Co., St. Louis, Mo.
UWANTA, Dittmann Shoe Co., St. Louis, Mo.
VACATION, Geddes-Brown Shoe Co., Indianapolis, Ind.
VACATION, Peters Shoe Co., St. Louis, Mo.
VACATION, Roberts, Johnson & Rand Shoe Co., St. Louis,
    Mo.
VALLEY CITY, Herold Bersck Shoe Co., Grand Rapids,
    Mich.
VALOIS, Thomson-Crooker Shoe Co., Boston, Mass.
VALUE, Friedman-Shelby Shoe Co., St. Louis, Mo.
VARSITY, Peters Shoe Co., St. Louis, Mo.
VASSAR, Hamilton-Brown Shoe Co., St. Louis, Mo.
VASSAR, Dittmann Shoe Co., St. Louis, Mo.
VERA, Rice & Hutchins, Boston, Mass.
VERONA, Roberts, Johnson & Rand Shoe Co., St. Louis, Mo.
VICI KID MOKORN, Chas Case Shoe Co., Worcester, Mass.
VICTOR, Friedman-Shelby Shoe Co., St. Louis, Mo.
VICTOR, Buchanan-Lawrence Co., Joliet, Ill.
VICTOR, Dittmann Shoe Co., St. Louis, Mo.
VILLAGE SCHOOL SHOE, Barton Bros., Kansas City, Mo.
VIM, Rice & Hutchins, Boston, Mass.
VIM, Marion Shoe Co., Marion, Ind.
VIOLA, Daniel Green Felt Shoe Co., Dolgeville, N. Y.
VIRGINIA, Dittmann Shoe Co., St. Louis, Mo.
VOGUE, The Pingree Company, Detroit, Mich.
WABASH, Brown Shoe Co., St. Louis, Mo.
WAGE EARNER, Endicott, Johnson & Co., Endicott, N. Y.
WALDORF, Framingham Shoe Co., South Framingham, Mass,
WALDORF, Brown Shoe Co., St. Louis, Mo.
WALKRITE, J. P. Hartray Shoe Co., Chicago, Ill,

WALK-OVER, Geo. E. Keith & Co., Brockton, Mass.
WALTON SHOE, A. G. Walton & Co., Boston, Mass.
WARM, Friedman-Shelby Shoe Co., St. Louis, Mo.
WARMACK, Roberts, Johnson & Rand Shoe Co., St. Louis, Mo.
WARM WOOL WITHIN, Blum Shoe Mfg. Co., Danville, N. Y.
WARREN GREEN, Northwestern Shoe Co., Seattle, Wash.
WARWICK, A. Priesmeyer Shoe Co., Jefferson City, Mo.
WARWICK, The Cady-Ivison Shoe Co., Cleveland, Ohio.
WASHINGTON, Washington Shoe Co., Haverhill, Mass.
WASHINGTON CUSTOM, Dittmann Shoe Co., St. Louis, Mo.
WATCH US, Hamilton-Brown Shoe Co., St. Louis, Mo.
WATER KING, Rice & Hutchins, Boston, Mass.
WATERS, Friedman-Shelby Shoe Co., St. Louis, Mo.
"WATSONTOWN", Payne Shoe Co., Charleston, W. Va.
WATSON SHOE, Watson Plummer Shoe Co., Chicago, Ill.
WAUKERZ, Walker & Whitman Campello, Mass.
WAVERLY, A. Priesmeyer Shoe Co., Jefferson City, Mo.
WAY DOWN EAST, Sawyer Boot & Shoe Co., Bangor, Me.
WAYNE, MADE, Wayne Shoe Mfg. Co., Ft. Wayne, Ind.
WEARBEST, Nolan-Earl Shoe Co., San Francisco, California.
WEARWEL TURN COMFORT, Lynch & Sherman, Lynn, Mass.
WEATHERBIRD, .Peters Shoe Co., St. Louis, Mo.
"WEBER," Weber Bros. Shoe Co., North Adams, Mass.
WEBSTER SCHOOL SHOES, Wertheimer-Swarts Shoe Co., St. Louis, Mo.
WEEWEAR, Guthmann, Carpenter & Telling, Chicago, Ill.
WELLSLEY, Banner Shoe Company, St. Louis, Mo.
WESTERN, Hamilton-Brown Shoe Co., St. Louis, Mo.
WESTERN LADY, A. Priesmeyer Shoe Co., Jefferson City, Mo.
WESTERN RESERVE, The Henry O. Werner Co., Columbus, Ohio.
WESTERN SHOP MADE, Friedman-Shelby Shoe Co., St. Louis, Mo.
WESTERN STAR, Brown Shoe Co., St. Louis, Mo.
WESTERN SCHOOL, A. Priesmeyer Shoe Co., Jefferson City, Mo.
WHIRL OF THE TOWN, Rice & Hutchins, Boston, Mass.
WHITE DONKEY, Friedman-Shelby Shoe Co., St. Louis, Mo.
WHITE HOUSE, Brown Shoe Co., St. Louis, Mo.
WHITE OAK, American Hand Sewed Shoe Co., Omaha, Neb.
WHITE PLUME. Friedman-Shelby Shoe Co., St. Louis, Mo.
WHITMAN SPECIAL, Walker & Whitman, Campello, Mass.
WHITTINGHILL, Whittinghill-Harlow Shoe Co., St. Joseph, Mo.

WHITTINGHILL-HARLOW, Whittinghall-Harlow Shoe Co., St. Joseph, Mo.
WICHERT & GARDINER, Wichert & Gardiner, Brooklyn, N. Y.
WICKLIFFE, The Cady-Ivison Shoe Co., Cleveland, Ohio.
WILLIAMS OUTING, A. A. Williams Shoe Co., Holliston, Mass.
"WILLIAMS MOLDER", A. A. Williams Shoe Co., Holliston, Mass.
WILL WEAR, Friedman-Shelby Shoe Co., St. Louis, Mo.
WINNA, Monadnock Shoe Co., Keene, N. H.
WINNER, Dittmann Shoe Co., St. Louis, Mo.
WINNER, Peters Shoe Co., St. Louis, Mo.
"WINNIE DAVIS", J. K. Orr Shoe Co., Atlanta, Ga.
WINTER, Peters Shoe Co., St. Louis, Mo.
WINTER, Endicott, Johnson & Co., Endicott, N. Y.
WIRE AND WELT, Barker, Brown & Co., Huntington, Ind.
WISCO ATHLETIC SHOES, Wisconsin Shoe Co., Milwaukee, Wis.
WISH BONE, Arnold-Henegar-Doyle Co., Knoxville, Tenn.
"WITCH ELK", Payne Shoe Co., Charleston, W. Va.
Wolfe's Foot Form, Wolfe Shoe Mfg. Co., Allentown, Pa.
WONDER, Peters Shoe Co., St. Louis, Mo.
WONDER, The Jos. P. Dunn Shoe & Leather Co., Denver, Colo.
WORCESTER, Peters Shoe Co., St. Louis , Mo.
"WORKIN COMFORT", J. K. Orr Shoe Co., Atlanta, Ga.
W. O. W., Burrow, Jones & Dyer Shoe Co., St. Louis, Mo.
Y. A., A. Priesmeyer Shoe Co., Jefferson City, Mo.
YALE, Dougherty-Fithian Shoe Co., Portland, Oregon.
YALE, A. Priesmeyer Shoe Co., Jefferson City, Mo.
YALE, Endicott, Johnson & Co., Endicott, N. Y.
YARDWIDE, Friedman-Shelby Shoe Co., St. Louis, Mo.
YATER, Burrow, Jones & Dyer Shoe Co., St. Louis, Mo.
YE OLD TOWN PUMP, The Plaut-Butler Co., Cincinnati, Ohio.
YE OLD TYME COMFORT SHOE, Linn & Sweet Shoe Co., Auburn, Maine.
YERMA CUSHION, F. Mayer Boot & Shoe Co., Milwaukee, Wis.
YOUNG AMERICA, I. P. Farnum, Chicago, Ill.
ZAIDEE, Dittmann Shoe Co., St. Louis, Mo.
ZERO, Hamilton-Brown Shoe Co., St. Louis, Mo.
18 CARAT, Peters Shoe Co., St. Louis, Mo.

List of Shoe Manufacturers

Twelve Hundred American Manufacturers with Address of Headquarters.

A.

Aborn, C. H. & Co., Lynn, Mass.
R. Ackerman Shoe Mfg. Co., Milwaukee, Wis.
Adams, Elihu T., Newburyport, Mass.
Adams Bros., Pittsfield, N. H.
Adams Shoe Co., Adamsdale, Pa.
Adams & Co., Hammond, La.
Adler, Joseph, Brooklyn, N. Y.
Adler, Martin & Katz, Rochester, N. Y.
Ahrens, Chas. A. & Co., Berlin, Ont., Can.
Albert, J. & Son, Brooklyn, N. Y.
Albright, H. S. & Co., Landingville and Orwigsburg, Pa.
Alden, C. H. Shoe Co., Abington, Mass.
Alden, Walker & Wilde, East Weymouth, Mass.
Aldrich, C. E. & Co., Farmington, N. H.
Alexander, Solomon, New York City.
Allen & Co., Philadelphia, Pa.
Allerton Bros., Seattle, Wash.
Allentown Shoe Mfg. Co., Allentown, Pa.
Allen, Foster-Willett Co., Lynn, Mass.
Alley, A. S. Shoe Co., Lynn, Mass.
All Wear Shoe Co., Catawissa, Pa.
Altschul, Julius, Brooklyn, N. Y.
Amazeen Shoe Co., Milwaukee, Wis.
American Shoe Mfg. Co., Allentown, Pa.
American Slipper Co., Philadelphia, Pa.
American Specialty Shoe Co., Milwaukee, Wis.
Ames-Holden, Ltd., Montreal and St. Hyacinthe, Que., Can.
Amherst Boot & Shoe Co., Ltd., Amherst, Nova Scotia.
Andrews, Wasgatt Co., Everett, Mass.
Anglemaire, Walter, Rockford, Ill.
Annapolis Larrigan Co., Annapolis, Nova Scotia.
Annar Shoe Mfg. Co., Philadelphia, Pa.
Apex Shoe Factory, New Orleans, La.
Apthorp Shoe Works, Littleton, N. H.
Archibold Bros., Ltd., Harbor Grace, Newfoundland.
Argus, Joseph, Buffalo, N. Y.
Armstrong, D. & Co., Rochester, N. Y.

Arnold, M. N. & Co., North Arlington. Mass.
Arrah Wanna Moccasin Co., Brockton, Mass.
Ashe, Noyes & Small Co., Auburn, Me.
Asher, L., Chicago, Ill.
Ashland Boot & Shoe Co., Ashland, Pa.
Ashuelot Shoe Co., Keene, N. H.
Athletic Shoe Co., Chicago.
Atlanta Leather Co., Atlanta, Ga.
Aylmer Shoe Co., Aylmer, Ont.

B.

B. & E. Shoe Co., Denver, Pa.
Baby Shoe Co., Dalmatia, Pa.
Baby Shoe Co., Newburyport, Mass.
Backaus, C. A. & Son, New Orleans, La.
Badger State Shoe Co., Madison, Wis.
Baker, Geo. W. Shoe Co., Brooklyn, N. Y.
Baker, George & Sons, Brooklyn, N. Y.
Baker, J. H. & Co., Beverly, Mass.
Bangor Shoe Mfg. Co., Bangor, Me.
Banister, J. A. C., Newark, N. J.
Barke & Co., Philadelphia, Pa.
Barker, Brown & Co., Huntington, Ind.
Barker, F. W., Georgetown, Mass.
Barber, Frank Shoe Co., N. Y. City, N. Y.
Barnard, J. W. & Son, Andover, Mass.
Barry, T. D. & Co., Brockton, Mass.
Barry, Wm. J., Salem, Mass.
Barry, Wm. J., Salem, Mass.
Bartels & Phelen Co., Chelsea, Mass.
Bartlett, James C., Haverhill, Mass.
Barton Bros., Kansas City, Mo.
Bass, G. H. & Co., Wilton, Me.
Bates, A. J. & Co., Webster, Mass.
Bates, Chas. S., Exeter, N. H.
Battreall, Whittinghill Shoe Co., St. Joseph, Mo.
Baumgaertel, Emil, Buffalo, N. Y.
Beal, The R. M. Leather Co., Lindsay, Ont., Can.
Beal Bros., Toronto, Ont.
Beals & Torrey Shoe Co., Milwaukee and Watertown, Wis.
Bean, A. J. & Son, Ottsville, Pa.
Bean Shoe Co., Salem, Mass.
Beardsley, Warren, Haverhill, Mass.
Beattie, R. L., Seattle, Wash.
Beatty Felting Co., Mishawaka, Ind.
Bebber Slipper Mfg. Co., Chicago, Ill.
Bedard, Louis, Ancienne Lorette, Que., Can.

Bedford Shoe Co., Carlisle, Pa.
Beeman Shoe Co., Dallas, Tex.
Belcher & Reed, Holbrook, Mass.
Bell Bros., Salem, Mass.
Bell, E. F. & Co., Beverly, Mass.
Bell, Geo. W., Pleasantville, N. Y.
Bell, J. & T., Montreal, Que., Can.
Bell, Wm. H., Pleasantville, N. Y.
Belleville Shoe Co., Belleville, Ill.
Belonga, George W. Co., Lynn, Mass.
Belwood Shoe Co., Rutherford College, N. C.
Benedict & Co., New Canaan, Conn.
Bennett, C. W. & Co., Fitchburg, Mass.
Benson & Holden, Auburn, N. Y.
Bentley & Olmstead, Des Moines, Iowa.
Beaumont & Traub, Lynn, Mass.
Bergmann, Theo. Mfg. Co., Portland, Ore.
Bering Felt Foot Co., Ltd., Berlin, Ont., Can.
Berlin Shoe & Slipper Co., Berlin, Ont.
Bernaque & Baillargeon, Three Rivers, Que., Can.
Bernstein, Wm., Brooklyn, N. Y.
Berry, A. H. Shoe Co., Portland, Me.
Bickford & Sweet, Worcester, Mass.
Bielefeld & Sahn, New York, N. Y.
Birdsboro Shoe Mfg. Co., Birdsboro, Pa.
Blake, Charles E. & Co., Lynn, Mass.
Blake, Allen Co., Beverly, Mass.
Bliss, E. M., Worcester, Mass
Bliss & Perry Co., Newburyport, Mass.
Blouin, J. B. & Son, Levis, Que., Can.
Bluff City Shoe Co., Hannibal, Mo.
Blum Shoe Mfg. Co., Dansville, N. Y.
Blyn Shoe Mfg. Co., New York, N. Y.
Blyn, Simon, New York, N. Y.
Bolander & Beckman, Lynn, Mass.
Bornhorst, P., Minster, O.
Borchardt, S. A. Co., N. Y. City, N. Y.
Bottomley, E. & Co., Haverhill, Mass.
Boyden Shoe Co., Newark, N. J.
Boyer, Xavier, Montreal, Que., Can.
Boyer, J. W. & Co., Victoria Corner, New Brunswick, Can.
Bowdoinham Shoe Co., Bowdoinham, Me.
Bradford Shoe Co., Columbus, Ohio.
Bradley & Metcalf Co., Milwaukee, Wis.
Brandau, A., Detroit, Mich.
Brandon Shoe Co., Aylmer, Ont., Can.
Braitling, F. K., Bridgeport, Conn.
Bray & Stanley, Beverly, Mass.

Brennen Boot & Shoe Co., Natick, Mass.
Brennan, J. M. Tannery & Shoe Factory, Upland, Cal.
Brennen & White, Brooklyn, N. Y.
Brett, C. M. Co., Hudson, Mass.
Brigham, F. & Gregory Co., Hudson, Mass.
Brigham, H. E., Westboro, Mass.
Broadwalk Shoe Co., Haverhill, Mass.
Brockton Co-operative B. & S. Co., Campello, Mass
Brockton Ideal Shoe Co., Brockton, Mass.
Brockton National Shoe Co., Brockton, Mass.
Brophy Bros. Shoe Co., Lynn, Mass.
Brown, A. E. & Co., Orwigsburg, Pa.
Brown, Carson, Schlefle Mfg. Co., Franklin, O.
Brown, Emery, Marblehead, Mass.
Brown, H. H. & Co., North Brookfield, Mass.
Brown, J. & Sons, Salem, Mass.
Brown Shoe Co., St. Louis and Moberly, Mo., and Murphysboro, Ill.
Buck, E. A. & Co., Bangor, Me.
Buckingham & Hecht, San Francisco, Cal.
Buckley Shoe Co., Houston, Tex.
Budd, J. F. Shoe Co., Burlington, N. J.
Buek & Co., Philadelphia, Pa.
Buffalo Shoe Co., E. Pepperell, Mass.
Burley & Stevens, Newburyport, Mass.
Burlington Shoe Co., Mt. Holly, N. J.
Burnham, C. A., Lynn, Mass.
Burrow, Jones & Dyer Co., St. Louis, Mo.
Burt, Edwin C. Co., Brooklyn, N. Y.
Burt, E. W. & Co., Lynn Mass.
Burt, E. W. Co., Lynn, Mass.
Butcher, W. A., Camden, N. J.

C.

C. & E. Shoe Co., Columbus, O.
Cahill Shoe Co., The, Cincinnati, O.
Cahn, Nickelsburg Co., Oakland, Cal., San Francisco, Cal.
Callahan & Myers, Allentown, Pa.
Campbell J. H., Lynn, Mass.
Capen Shoe Co., Williamette Falls, Ore.
Cahart Shoe Mfg. Co., Atlanta, Ga.
Carr, Chas. G., Brooklyn, N. Y.
Carleton, George F. & Co., Haverhill, Mass.
Carleton and Hunt, Haverhill, Mass.
Carlisle Shoe Co., Carlisle, Pa.
Carruthers-Jones Shoe Co., St. Louis, Mo.
Carter, J. W. Shoe Co., Beverly, Mass.

Carter, J. W. & Co., Nashville, Tenn.
Case, Chas. Shoe Co., Worcester, Mass.
Case, W. D., Putnam, Conn.
Cass & Daley, Salem, Mass.
Cate-Quimby Shoe Co., Northwood Narrows, N. H.
Century Shoe Co., Macungle, Pa.
Chadwick, J. H. P. & Co., Marblehead, Mass.
Chambersburg Shoe Mfg. Co., Chambersburg, Pa.
Chandler, D. Harry, Vineland, N. J.
Chaplin, G. W. & Co., Georgetown, Mass.
Charboneau, F. X., Montreal, Que., Can.
Charter Oak Shoe Co., Wethersfield, Conn.
Chase, Chamberlain & Co., Raymond, N. H.
Chase, F. S., Haverhill, Mass.
Chase, W. F., Haverhill, Mass.
Chase, W. S. & Sons, Haverhill, Mass.
Chesley & Rigg, Haverhill, Mass.
Chippewa Shoe Mfg. Co., Chippewa Falls, Wis.
Churchill & Alden Co., Campello, Mass.
Cimon, A. P. Shoe Mfg. Co. Montreal, Que., Can.
Cincinnati Shoe Co., Bethel, Ohio.
City Made Slipper Co., New York.
Clairoux & Richer, Montreal, Que., Can.
Clapp, Edwin & Son, East Weymouth, Mass.
Clapp & Tapley, Danvers, (Tapleyville), Mass.
Clark, John T. & Co., Bangor, Me.
Clark, W. P. & Co., Marblehead.
Clement & Ball Shoe Mfg. Co., Baltimore, Md.
Clifford, E. J. & Co., Haverhill, Mass.
Clogston, H. W., Haverhill, Mass.
Cloutman, J. F. & Co., Farmington, N. H.
Cobb, Chas. H., Lynn, Mass.
Cogan, P. & Son, Stoneham, Mass.
Gogswell Shoe Co., Haverhill, Mass.
Cogswell, James, Lynn, Mass.
Cohen & Frank Co., Brooklyn, N. Y.
Cohn, Jessie Shoe Co., Chicago, Ill.
Cole, B. E. & Co., Newburyport, Mass.
Collins, C. M., S. Danville, N. H.
Collins, H. S., Haverhill, Mass.
Collins, L., Waldo, Kingston, N. H.
Colmary, A. H. & Co., Baltimore, Md.
Columbia Shoe Co., Richmond, Va.
Columbia Shoe Co., Columbia, Pa.
Columbia Shoe Co., Sheboygan, Wis.
Comfort Shoe Co., Albany, N.Y.
Comfort Slipper Co., New York, N. Y.

51

Commonwealth Shoe & Leather Co., Whitman, Mass., and
Gardiner and Showkegan, Me.
Comtois, J. E., Laurierville, Que., Can.
Condon Bros. & Co., Brockton, Mass.
Connell, J. & Co., Lynn, Mass.
Connelly, J. J. Co., Salem, Mass.
Connelly Shoe Co., Stillwater, Minn.
Conrad Shoe Co., Louisville, Ky.
Cook-Fitzgerald Co., Ltd., London, Ont., Can.
Cook, J. A. & Bro., Lynn, Mass.
Cooper, A. R., Findlay, O.
Cooper, Dunn & Co., Pontiac, Ill.
Copeland, Ellis F. & Son, Brockton, Mass.
Copeland & Ryder Co., Jefferson, Wis.
Corbell, A., Montreal, Que., Can.
Corbin, B. A. & Son Co., Webster, Mass.
Corcoran, Timothy, Brockton, Mass.
Cort, Chas., Newark, N. J.
Cote, J. & M., La Compagnie, St. Hyacinthe, Que., Can.
Cotter Shoe Co., Lynn, Mass.
Cotton, W. O., Philadelphia, Pa.
Cousins, J. & T., Brooklyn, N. Y.
Craddock-Terry Co., Lynchburg, Va.
Crafts, G. P. Co., Manchester, N. H.
Cramer, J. & Son, Brooklyn, N. Y.
Crapault, B., Quebec, Que., Can
Creighton, A. M., Lynn, Mass.
Crescent Shoe Co., Reading, Pa.
Criterion Shoe Co., Beverly, Mass.
Cross, John, H., Cambridge, Mass.
Crosby, The H. H. Co., Ltd., Hebron, Nova Scotia.
Crossett, Lewis A. Co., N. Abington Mass.
Crotty & Gardiner Brooklyn, N. Y.
Crown Shoe Co., Boston, Mass.
Croxton, Wood & Co., Philadelphia, Pa.
Cummings, David Co., S. Berwick, Me.
Cummings, The Co., Worcester, Mass.
Currier, Andrew, Newton, N. H.
Currier, T. J., Lynn, Mass.
Curtis & Jones Co., Reading, Pa.
Cushman, John S., Lynn, Mass.
Cushman & Hebert, Haverhill, Mass.
Cushman-Hollis Co., Auburn, Me.
Cutter, A. A. Co., Eau Claire, Wis.
Czerney, W. J., Brooklyn, N. Y.

D.

Dack, R. & Son, Toronto, Ont., Can.

Daltz, George, Rochester, N. Y.
Damon & Ellis, Boston, Mass.
Daniels, Geo. F. Co., Lynn, Mass.
Daoust, Lalonde & Co., Montreal, Que., Can.
Davis, H. E. Co., Freeport, Me.
Davis Shoe Co., Lynn, Mass.
Dayfoot, C. B. & Co., Georgetown, Ont., Can.
Dayton, J. E., Co., Williamsport, Pa.
Dean, Charles W. & Co., Cochituate, Mass.
Delavan Shoe Co., Delavan, Wis.
Delaware River Shoe Mfg. Co., Beverly, N. J.
Delta Shoe Co., Cambridge, Mass.
Dengler Bros., Pottsville, Pa.
Derry Shoe Co., Derry, N. H.
Derry & Wilkins, Lynn, Mass.
Desmond-Hayden Shoe Co., Lynn, Mass.
Desnoyers Shoe Co., Springfield, Ill., and St. Louis.
Dessner, Jacob, Brooklyn, N. Y.
Devine & Yungel, Harrisburg, Pa.
De Wolfe Shoe Co., Conway, Mass.
Dibble, W. B., Lynn, Mass.
Diggs-Vanneman Mfg. Co., Baltimore, Md.
Dingley, Foss Shoe Co., Auburn, Me.
Disautels, Anable & Co., Montreal, Que., Can.
Dion, D. & Co., Quebec, Que., Can.
Dittmann, Geo. F. Boot & Shoe Co., St. Louis, Mo.
Divac, Sharp & Co., Philadelphia, Pa.
Dickinson Shoe Co., Lynn, Mass.
Dix, Roberts Shoe Mfg. Co., Brooklyn, N. Y.
Dixon-Bartlett Co., Baltimore, Md.
Dodge, Arthur F., Beverly, Mass.
Dodge, Nathan D. & Sons, Newburyport, Mass.
Dodd, Jos. M., Brooklyn, N. Y.
Dolge, Alfred Mfg. Co., Dolgeville, Cal.
Dolgeville Felt Shoe Co., Dolgeville, N. Y.
Dollison, J. M. & Co., Chicago, Ill.
Dooley, M. J., Ashley, Pa.
Donovan, D. A. & Co., Lynn, Mass.
Donovan, James Co., Everett, Mass.
Donovan, James P., Brockton, Mass.
Donovan, John R., Lynn, Mass.
Dorell, W. H. & Son, Canden, N. J.
Dorsch, Wm. & Sons Mfg. Co., Newark, N. J.
Douglas, W. L. Shoe Co., Brockton, Mass.
Dow, W. L., Haverhill, Mass.
Dowd, C. E. & T. F., Natick, Mass.
Drew, Irving Co., Portsmouth, Ohio.
Drew, W E. & Co., Richmond, Va.

Dreyer, John H. & Co., Baltimore, Md.
Drolet, J. B. & Co., Quebec, Que., Can.
Duchaine Shoe Co., Quebec, Que., Can.
Dube, J. A. & Co., Montreal, Que., Can.
Dudley, L. B. & Co., Haverhill, Mass.
Dufresne & Locke, Maisonneuve, Que., Can.
Duhig & Co., Brooklyn, N. Y.
Dugan & Hudson Co., Rochester, N. Y.
Durgin Shoe Co., Haverhill, Mass.
Dunn & McCarthy, Auburn and Binghamton, N. Y.
Dupont & Frere, Montreal, Que., Can.
Durland-Weston Shoe Co., Honesdale, Pa.
Durkle, A. A., Yarmouth, N. S.
Durling, E. J., Haverhill, Mass.
Duttenhofer, Val Sons Co., Cincinnati, O.
Dyer, Mortimer Co., Pittsfield, N. H.

E.

Eady Shoe Co., Otsego and Hopkins, Mich.
Eagle Shoe Co., Montreal, Que., Can.
Eagle Shoe Mfg. Co., Lynn, Mass.
Eastman Shoe Co., Lynn, Mass.
Eaton, Charles A. Co., Augusta, Me., and Brockton, Mass.
Ebbets, John Shoe Co., Buffalo, N. Y.
Eby Shoe Co., Lititz, Pa.
Edwards, J. & Co., Philadelphia, Pa.
Eisenhuth, T. H. & Co., Sellins Grove, Pa., and Williamsport, Pa.
Elkin Shoe Co., Elkin, N. C.
Elkin, M. & Co., Philadelphia, Pa.
Elkskin Shoe Co., Los Angeles, Cal.
Elkskin Moccasin Mfg. Co., Ypsilanti, Mich.
Emerson, C. S. & Co., Derry, N. H.
Emerson & Wasson, Haverhill, Mass.
Emerson Shoe Co., Rockland, Mass.
Emerson's, W. A. Sons, Hampstead, N. H.
Emery, C. P. & Co., Haverhill, Mass.
Emery & Marshall, Haverhill, Mass.
Endicott-Johnson Co., Lestershire & Endicott, N. Y.
Engel-Cone Shoe Co., E. Boston, Mass.
Ennis, John, Brooklyn, N. Y.
Enterprise Shoe Mfg. Co., Brooklyn, N. Y.
Estabrook-Anderson Shoe Co., Nashua, N. H.
Evans, L. B. Son Co., Wakefield, Mass.
Evans Shoe Co., Napa, Cal.
Evans Bros. Shoe Co., West Newbury, Mass.
Excelsior Shoe Co., Portsmouth and Ironton, O.
Excelsior Shoe Co., Brooklyn, N. Y.

Excelsior Shoe & Slipper Co., Cedarburg, Wis.

F.

Fairfield Shoe Co., Lancaster, O.
Falconer & Feeley, Raymond, N. H.
Fall Creek Mfg. Co., The, Freeville, N. Y.
Fallek, S., St. Louis, Mo.
Fargo & Phelps, Chicago, Ill., and Louisiana, Mo.
Fargo Shoe Mfg. Co., Belding, Mich.
Farmer Shoe Co., Acton Vale, Que., Can.
Farmington Shoe Mfg. Co., Farmington, N. H.
Faunce & Spinney, Lynn, Mass.
Federal Shoe Co., Lowell, Mass.
Felch, W. L. & Co., Natick, Mass. ·
Felder Shoe Mfg. Co., Seattle, Wash.
Felter & Co., Newark, N. J.
Feldner, P. Shoe Co., St. Cloud, Wis.
Fenerty, Cassoboon & May, Lynn, Mass
Ferndale Shoe Mfg. Co., Ferndale, Pa.
Ferris, Isaac, Jr., Co., Camden, N. J.
Fiebrich-Fox-Hilker Shoe Co., Racine, Wis.
Field, Lumbert Co., Brockton, Mass.
Field, Fred F. Co., Brockton, Mass., and Providence, R. I.
Field, P. A., Shoe Co., Salem, Mass.
Field Bros. & Gross Co., Auburn, Me.
Finch Bros., Marblehead, Mass.
Finch Shoe Co., Springfield, O.
Fisher, Alfred D., Lynn, Mass.
Fisher, A. & Son, Lynn, Mass.
Fiske Shoe & Leather Co., Holbrook, Mass.
Fitzgerald, Thomas, Brooklyn, N. Y.
Fitzpatrick, Frank L., Philadelphia, Pa.
Florsheim Co., Chicago, Ill.
Foot, Schulze & Co., St. Paul, Minn.
Foot, S. B. & Co., Red Wing, Minn.
Foote, L. R., Rochester, N. Y.
Foot-Ease Shoe Co., Haverhill, Mass.
Forbush Shoe Co., North Grafton, Mass.
Ford, C. P. & Co., Rochester, N. Y.
Ford, H. R. & Co., Lynn, Mass.
Foss, Packard & Co., Auburn, Me.
Foster, A. J., Lowell, Mass.
Foster, John F. & S., Avon, Mass.
Foster, John Co., Beloit, Wis.
Foster, Moulton Shoe Co., Brookfield, Mass.
Foster, O. I. & Co., Haverhill, Mass.
Foster's Wm. C. Sons, Rowley, Mass.

Fox, Chas. K., Haverhill, Mass., & Wolfboro, N. H.
Fox, David, Wilmington, Del.
Fox, F. J., Rochester, N. Y.
Framingham Shoe Mfg. Co., South Framingham, Mass.
Freeland, H. H. Rochester, N. Y.
Freed Bros., Richlandtown, Pa.
Freeport Shoe Mfg. Co., Freeport, Ill.
French, J. E. & Co., Rockland. Mass.
French, J. F., Haverhill, Mass.
French, Shriner & Urner, Boston, Mass.
Fress, Geo. J., Camden, N. J.
Friedman-Shelby Shoe Co., St. Louis, and Kirksville and
    Mexico, Mo.
Friend Shoe Mfg. Co., Petersburg, Mass.
Frye, John A. Shoe Co., Marlboro, Mass.
Fuhrer, Henry, Baltimore Md.
Fuller, Chandler & Patten Co., Hudson, Mass.
Furber, D. L., Dover, N. H.

———

G.

Gale Bros., Exeter, N. H.
Gale Bros., Quebec, Que., Can.
Gale Shoe Mfg. Co., Haverhill, Mass., and Portsmouth, N. H.
Gage & Russ, Haverhill, Mass.
Galt Shoe Mfg. Co., Galt, Ont., Can.
Garden City Shoe Co., Beverly, Mass.
Garside, A. & Sons, New York, N. Y.
Gauthier, The Louis Co., Quebec, Que., Can.
Georgetown B. & S. Co., Georgetown, Mass.
Gerber, E. C. Shoe Co., Orwigsburg, Pa.
Germuth, A. & Son, New York City.
Getchell, M. L. & Co., Monmouth, Me
Getty & Scott, Ltd., Calt, Ont., Can.
Getz, A., Shoe Mfg. Co., Lancaster, O.
Gibbon, Chas. S. Co., Philadelphia, Pa
Gibbs, O. A. Shoe Co., Dover, N. H.
Gilchrist, J. P., Rochester, N. Y.
Glen Mfg. Co., Harrisburg, Pa.
Gloyd, A. E., Shoe Co., Lynn, Mass.
Glover, Daniel & Co., Salem, Mass.
Godman, H. C. Co., Columbus, O.
Gokey, W. N. Shoe Co., Jamestown, N. Y.
Goldberg, Henry, Philadelphia, Pa.
Golden Sporting Shoe Co., Brockton, Mass.
Goller Shoe Co., Lynn, Mass.
Goodrich, Hazen B. & Co., Haverhill, Mass.
Goodwin, J. J., Rochester, N. Y.
Gorman, N. F., Haverhill, Mass.

Gotzian, C. & Co., Chippewa Falls and Eau Claire and St.
Paul, Minn.
Gould, G. A., Topsfield, Mass.
Goulet, Quebec, Que., Can.
Graham, Bumgarner Co., Parkersburg, W. Va.
Grant Shoe Co., Lynn, Mass.
Gray's H. W. Sons, Syracuse, N. Y.
Greeley, A. W., Haverhill, Mass.
Green, C. E. & Co., Manchester, N. H.
Green, Daniel, Felt Shoe Co., Dolgeville, N. Y.
Greenberg, I. & Co., New York, City.
Greenberg & Miller, New York City.
Green-Wheeler Shoe Co., Fort Dodge, Ia.
Greenville Shoe Co., Greenville, Ill.
Gregory, F. E. Co., Lynn, Mass.
Greilich, Wm., Brooklyn, N. Y.
Griffin-White Shoe Co., Brooklyn, N. Y.
Griffin, W. H., Manchester, N. H.
Griffith, C. D. Shoe Co., Denver, Col.
Grosch, J. G. Felt Shoe Co., Milverton, Ont., Can.
Grossman, Julius, New York, N. Y.
Grosz, F. A., New Orleans, La.
Grover's J. J. Sons, Lynn, Mass.
Guptill, H. E., Haverhill, Mass.

H.

Hagerty, P. Shoe Co., Washington C. H., Ohio.
Halifax Shoe Co., Halifax, Pa.
Hagel, A. & E., Philadelphia, Pa.
Hallahan & Sons, Philadelphia, Pa.
Hamburg Felt Boot Co., New Hamburg, Ont., Can.
Hamilton-Brown Shoe Co., St. Louis and Columbia, Mo.
Hamilton, W. B. Shoe Co., Toronto, Ont., Can.
Hammen, F. J., Utica, N. Y.
Hammonton Shoe Co., Hammonton, N. J.
Hanan & Son, Brooklyn, N. Y.
Hand Made Shoe Co., Chippewa Falls, Wis.
Hannibal Shoe Co., Hannibal, Mo.
Hanson, J. M., San Jose, Cal.
Harbor Grace Boot & Shoe Mfg. Co., Ltd., Harbor Grace,
Newfoundland.
Harding Cushion Shoe Co., Boston, Mass.
Harding J. Jordan Shoe Co., New Sharon, Me.
Harney Bros., Lynn, Mass.
Harney, P. J. Shoe Co., Lynn, Mass.
Harnishfeger Shoe Mfg. Co., Evansville, Ind.
Harsh & Edmonds Shoe Company, Milwaukee, Wis.

Hartt Boot & Shoe Co., The, Ltd., Fredericton, New Brunswick, Can.
Hartford Slipper Mfg. Co., Hartford, Conn.
Harris Shoe Co., Haverhill, Mass.
Harrisburg Shoe Mfg. Co., Harrisburg, Pa.
Haskins Shoe Mfg. Co., Stittville, N. J.
Hawkes, G. A. Co., Richmond, Me.
Hawkins Sons, East Norwalk, Conn.
Hayden & Collins, Haverhill, Mass.
Hayes, E., New York City.
Hazzard, R. P. Co., Gardiner, Me.
Healy Bros. Shoe Co., Stoneman, Mass.
Heath, I. L., Rochester, N. Y.
Hebard, T. C. & Co., Windsor, Vt.
Hedlund Shoe Co., Salem, N. H.
Helmers, Bettman & Co., Cincinnati, O.
Helming-McKenzie Shoe Co., Cincinnati, O.
Henderson, The J. S. Co., Ltd., Parrsboro, Nova Scotia.
Henne, Wm. & Co., Brooklyn, N. Y.
Hennessey Shoe Mfg. Co., Cincinnati, O.
Herrman, Jandorf & Oxberry Co., New York, N. Y.
Herman, Joseph M. & Co., Millis, Mass.
Herold-Bertsch Shoe Co., Grand Rapids, Mich.
Herrick, G. W. Shoe Co., Lynn, Mass.
Hess, N. & Bro., Baltimore, Md.
Hewston, J. W. Co., Ltd., Toronto, Ont., Can.
Heywood Boot & Shoe Co., Worcester, Mass.
Higgins, L. & Co., Yarmouth, Nova Scotia.
Hillsdale Shoe Co., Hillsdale, Mich.
Hilliard & Tabor, Haverhill, Mass.
Hine & Lynch, Poughkeepsie, N. Y.
Hirschfield, M. R., Maysville, Ky.
Hirth, Krause Co., Rockford, Mich.
Hitchcock & Allard, Haverhill, Mass.
Hoag & Walden, Lynn, Mass.
Hodgdon, F. M., Haverhill, Mass.
Hodous, J. S., Cleveland, O.
Hodsdon Mfg. Co., Yarmouthville, Me.
Hoff, M. H. & Co., Natick, Mass.
Hogan Shoe Co., Cincinnati, Ohio, and Aurora, Ind.
Hoge-Montgomery Co., Frankfort, Ky.
Holland Shoe Co., Holland, Mich.
Holters-Cravin Co., Cincinnati, O.
Holmes, F. B. Co., Chelsea, Mass.
Holt, Moses K., Haverhill, Mass.
Honesdale Footwear Co., Honesdale, Pa.
Honesdale Shoe Co., Honesdale, Pa.
Honesdale Union Stamp Shoe Co., Honesdale, Pa.

Hopkins Glove Co., Cincinnati, O.
Hopkins, J. T. & Sons, Salem, Mass.
Houghton, Warren Co., Somersworth, N. H.
How, C. M., Haverhill, Mass.
How, Geo. C., Haverhill, Mass.
Howard, Briggs & Fray Co., Auburin, Me.
Howard & Foster, Brockton, Mass.
Howe Shoe Co., Lynn, Mass.
Howe, S. H. Shoe Co., Marlboro, Mass.
Hoyt, F. M. Shoe Co., Manchester, N. H.
Hoyt, F. M. Shoe Co., Manchester, N. H.
Hoyt, Frank, Lowell, Mass.
Hoyt, Rowe & Co., Lynn, Mass.
Hubler-Earnshaw Co., Philadelphia, Pa.
Huckins & Temple, Milford, Mass.
Huiskamp Bros., Co., Keokuk, Ia., and Warsaw, Ill.
Humphrey & Paine, Marblehead, Mass.
Hunkins, W. O., Haverhill, Mass.
Hurlbut Co., Preston, Ont.
Hurley Shoe Co., Rockland, Mass.
Hyde, A. R. Shoe Co., Somerville, Mass.
Hyer, C. H., Olathe, Kansas.
Hyman Bros., Rochester, N. Y.
Huyett & Rhoades, Birdsboro, Pa.

I.

Ingals, W. H., Lynn, Mass.
Ideal Baby Shoe Co., Danvers, Mass.
Ireland, Grafton Co., Dover, N. H.

J.

Jacob, H. & Sons, New York, N. Y.
Jacobi, Phillip, Toronto, Ont., Can.
Jacobs, Joseph Shoe Co., New York, N. Y.
Jefts, L. T. Co., Hudson, Mass.
Jenkins Bros. Shoe Co., Winston-Salem, N. C.
Jesseman, H. L. Co., Marblehead, Mass.
Jobin & Rochette, Quebec, Que., Can.
Johansen Bros. Shoe Co., St. Louis, Mo.
John & Beck Shoe Co., Jefferson, Wis.
Johnson Bros. Shoe Mfg. Co., Hallowell, Me.
Johnson Bros. & Co., Jamestown, N. C.
Johnson, C. W., Natick, Mass.
Johnson, Luther S. & Co., Lynn, Mass.
Johnson-Baillie Shoe Co., Millersburg, Pa.
Johnson Shoe Co., Jamestown, N. C.

Johnson, W. S. & Co., Putnam, Conn.
Johnston, C. K. & Co., Eureka, Cal.
Johnston & Murphy, Newark, N. J.
Jones, E. & Co., Spencer, Mass.
Jones, T. H. Shoe Co., Stoneham, Mass.
Jones, V. K. & A. H. Co., Lynn, Mass.
Jorolemon-Oliver & Co., Rochester, N. Y.
Joslin, A. L. Co., Oxford, Mass.
Julian & Kokenge Co., Cincinnati, O.
June, W. J. & Co., Rochester, N. Y.
Justin, H. J., Nocoma, Texas.

K.

Kalt-Zimmers Mfg. Co., Milwaukee, Wis.
Kaufman, Emil Co., New York, N. Y.
Keene, Jacob, Athens, Mich.
Keighley, Chas. & Sons, Vineland, N. J.
Keith, Geo. E. Co., Campello and North Adams and Middle-
    boro, Mass.
Keith & Pratt, Middleboro, Mass.
Keith, Preston B. Shoe Co., Brockton, Mass.
Kelley, Martin Co., Danvers, Mass.   ·
Kelly, M. E. Castleton, N. Y.
Kelly, Buckley Co., Brockton, Mass.
Kenmore Shoe Co., Williamsport, Pa.
Kenny, L. W. & Co., Lynn, Mass.
Kentucky Shoe Mfg. Co., Eddyville, Ky.
Kepner-Scott Shoe Co., Odwigsburg, Pa.
Keystone Shoe Co., Philadelphia, Pa.
Keystone Shoe Mfg. Co., Kutztown, Pa.
Kiely, T. J. & Co., Lynn, Mass.
Kimball, Alfred Shoe Co., Lawrence, Mass.
Kimball Bros. Shoe Co., Manchester, N. H.
Kimball, W. & V. O., Haverhill, Mass.
Kimball-Knight Shoe Co., Newburyport, Mass.
King, Mrs. A. R., Corporation, Lynn, Mass.
King, Richard E., Nyack, N. Y.
Kingsbury Footwear Co., Maisonneuve and Montreal, Que.,
    Can.
Kirkendall, F. P. & Co., Omaha, Neb.
Kirvan-Doig, Ltd., Montreal, Quebec, Can.
Knights and Perry, Haverhill, Mass.
Knipe Bros., Ward Hill, Mass.
Koeckert, Aug., Jr., Cleveland, O.
Koerner, Marsh Shoe Co., Milwaukee, Wis.
Kollock, F. A., Lynn, Mass.
Kozak & McLaughlin, New York.
Kozy Slipper Co., Lynn, Mass.

Kreider Shoe Mfg. Co., Elizabethtown, Pa., Annville, Palmyra.
Kreiders, W. L. Sons Mfg. Co., Palmyra, Pa.
Krieger & Co., Brooklyn, N. Y.
Krieger Shoe Co. (The), Brooklyn, N. Y.
Kringler, F. X., Philadelphia, Pa.
Krippendorf-Dittman Co., Cincinnati, Ohio.
A.-O. Shoe Co., Cincinnati, O.
Krohn, Fechheimer & Co., Cincinnati, O.
Krohngold, Max, Cleveland, O.
Kyng, B. Shoe Co., Rochester, N. Y.
Kuchman, Joseph, Rochester, N. Y.
Kutz, G. M. Shoe Co., San Francisco, Cal.

L.

L. & W. Shoe Co., Chippewa Falls, Wis.
Lachance & Tanguay, Quebec, Que., Can.
La Crosse Boot & Shoe Mfg. Co., La Crosse, Wis.
Laird, Schober & Co., Philadelphia, Pa.
Lancaster Shoe Co., Lancaster, O.
Lancy, John, Jr., Marblehead, Mass.
Landis, J. Shoe Co., Palmyra, Pa.
Langlois, J. S. & Co., Quebec, Que., Can.
Lane, William, Brooklyn, N. Y.
Lantzky-Allen Shoe Co., Dubuque, Iowa.
Larson, C. P. Shoe Co., Baraboo, Wis.
Latteman, John J. Shoe Mfg. Co., Brooklyn, N. Y.
Lawrence Clog Works, Lawrence, Mass.
Leach Shoe Co., Rochester, N. Y.
Learned, Geo. A. Co., Newburyport, Mass.
Leavitt, F. E., Haverhill, Mass.
Leavitt, G. B. & Co., Haverhill, Mass.
Leckie, J. Co., Ltd., Vancouver, British Columbia, Can.
Leech Bros., Riverside, N. J.
Leech Shoe Co., Burlington, N. J.
Lee, W. H., Memphis, Tenn.
Lee Bros. Co., Athol, Mass.
Lefavour, C. P. Shoe Co., Beverly, Mass.
Lefavour, D. D. & Co., Salem, Mass.
Legg, A. M. Shoe Co., Pontiac, Ill.
Lehigh Valley Shoe Co., Allentown, Pa.
Lemay, H., Milwaukee, Wis.
Lenox Shoe Co., Philadelphia, Pa.
Leonard, Shaw & Dean, Middleboro, Mass.
Leonard Shoe Co., Lynn, Mass.
Leonard & Barrows, Middleboro, Mass., and Belfast, Me.
Levi Shoe Co., Chicago, Ill.
Levirs & Sargent, Lynn, Mass.
Lewis, G. W. & Son, Burlington, N. J.

Lewis, Herman E., Haverhill, Mass.
Lewis, W. C. Co., Haverhill, Mass.
Liberman, S., Brooklyn, N. Y.
Lilliputian Shoe Co., Abington, Mass.
Lindenberg, H., New Orleans, La.
Linder Shoe Co., Carlisle, Pa.
Linscott-Tyler-Wilson Co., Rochester, N. H.
Linton, James & Co., Montreal, Que., Can.
Little, A. E. & Co., Lynn, Mass.
Little Falls Felt Shoe Co., Little Falls, N. Y.
Livingston, E. C., New Oxford, Pa.
Lloyd-Adams Co., The, Portsmouth, O.
Locke, J. I., Haverhill, Mass.
Logan, Thomas H. Co., Lynn, Mass.
Logan, Edward F., Lynn, Mass.
Lounsbury, Matthewson & Co., South Norwalk, Conn.
Lounsbury & Soule, Stamford, Conn.
Lowell Shoe Co., Lowell Mass.
Lowry & Son, Huntsville, Ala.
Luddy, James Shoe Co., Dover, N. H.
Lumberton Shoe Co., Lumberton, N. J.
Lunn & Sweet Shoe Co., Auburn, Me.
Lutz, C. A. & Co., Columbus, Neb.
Lynn, J. A., Haverhill, Mass.
Lynch & Sherman, Lynn, Mass.
Lyon, F. C., New York, N. Y.
Luscomb-Kadel Co., Port Jarvise, N. Y.

M.

Macdonald & Kiley Co., Cincinnati, O.
Macfarlane Shoe Co., Montreal, Que., Can.
Mackenzie, Crowe & Co., Ltd., Bridgetown, Nova Scotia.
Maetrich, Eyre & Co., Brooklyn, N. Y.
Maibach, Jacob, Philadelphia, Pa.
Malbon, M. D. Co., Haverhill, Mass.
Manier, Dunbar Co., Nashville, Tenn.
Manistee Shoe Mfg. Co., Manistee, Mich.
Mansfield, G. A. & E. A., Lynn, Mass.
Manss Shoe Mfg. Co., Cincinnati, O.
Marden, John E., Philadelphia, Pa.
Marion Shoe Co., Marion, Ind.
Marier & Trodel, Quebec, Can.
Marietta Shoe Co., Williamstown, W. Va.
Marks, L. V. & Co., Augusta, Ky., and Ripley, O.
Marks, Stix & Sacks Co., The, Cincinnati, O.
Marsh, Wm. & Shoe Co., Quebec, Que., Can.
Marshall, C. S. & Co., Brockton, Mass.
Marsters, J. A. & Co., Beverly, Mass.

Marston, A. F., Lynn, Mass.
Marston, C. S., Jr., Haverhill, Mass.
Martin, E. P., Marblehead, Mass.
Martin Slipper Co., Haverhill, Mass.
Marzluff, F. M. Co., Three Rivers, Que., Can.
Mason Shoe Mfg. Co., Chippewa Falls, Wis.
Mass, States Prison, Boston, Mass.
Masse, Joseph & Co., Three Rivers, Que., Can.
May, W. B., Bridgewater, Mass.
Mayer, F. Boot & Shoe Co., Milwaukee, Wis.
Maynard Shoe Co., Claremont, N. H.
McBrearty, John, Philadelphia, Pa.
McCord-Donovan Shoe Co., St. Joseph, Mo.
McCready, The James Co., Montreal, Que., Can.
McDermott Shoe Co., Montreal, Que., Can.
McDonald & Kiley Co., The, Cincinnati, O.
McElwain, W. H. Co., Boston and Bridgewater, Mass., Manchester and Newport, N. H.
McKeen, The C. E. Co., Quebec, Que., Can.
McLaughlin, W. P. & Co., Haverhill, Mass.
McMaster, J. J., Rochester, N. Y.
McNamara, L. F. Co., The, Haverhill, Mass.
McNamara, S. B. & Co., Haverhill, Mass.
McPherson, John Co., Ltd., Hamilton, Ont.
Medlar & Holmes Co., Philadelphia, Pa.
Mehringer, Gustaf, New York, N. Y.
Meier, John Shoe Co., St. Louis, Mo.
Melanson, J. I. & Bro., Lynn, Mass.
Meldola & Coon, Rochester, N. Y.
Mendenhall, S. H. & Co., High Point, N. C.
Menihan Co., Rochester, N. Y.
Menzies Shoe Co., Detroit, Mich.
Mercier & Co., Montreal, Quebec.
Merrill, A. J., Haverhill, Mass.
Merrill, Porter & Co., Lynn, Mass.
Merriam, H. W. Shoe Co., Newton, N. J.
Merritt, John R., Lynn, Mass.
Middlesex Shoe Co., New Brunswick, N. J.
Milford Shoe Co., Milford, Mass.
Millar & Wolfer, Chelsea, Mass.
Miller, A. M. & Co., Orwigsburg, Pa.
Miller, F. E., Rochester, N. Y.
Miller, I., New York, N. Y.
Miller, Hess & Co., Akron, Pa.
Miller, Joseph Co., Racine, Wis.
Miller Shoe Co., Montreal, Que., Can.
Miller Shoe Mfg. Co., Cincinnati, O.
Miller, W. Y., Schuylkill Haven, Pa.

Millett, Woodbury & Co., Beverly, Mass.
Milton Shoe Co., Milton, N. H.
Minister-Myles Shoe Co., Ltd., Toronto, Ont., Can.
Minnesota Shoe Co., St. Paul, Minn.
Minchin, Joseph, Brooklyn, N. Y.
Minor, P. W. & Son, Batavia, N. Y.
Minor-Pullen Co., Hightstown, N. J.
Mishawaka Woolen Mfg. Co., Mishawaka, Ind.
Missouri Slipper Co., St. Louis, Mo.
Mitchell, Frank, Marblehead, Mass.
Mitchell-Caunt, E. Lynn, Mass.
Model Shoe Co., St. Louis, Mo.
Modern Shoe Co., Pontiac, Ill.
Moloney Bros. Co., Rochester, N. Y.
Monadnock Shoe Co., Keene, N. H.
Monarch Shoe Co. (The), Racine, Wis.
Monnig Shoe Co., Marshall, Mo.
Montgomery & Co., Columbus, O.
Moody, Emerson Shoe Co., West Derby, N. H.
Moore-Shafer Shoe Mfg. Co., Brockport, N. Y.
Morgan Shoe Co., Newburyport, Mass.
Morr Shoe Mfg. Co., Ashland, O.
Motteler Shoe Co., Louisville, Ky.
Mountain State Shoe Co., Huntington, W. Va.
Muir, The James Co., Ltd., Quebec, Que., Can.
Mullen, J. D. & Son, Lynn, Mass.
Munster, A., Dallas, Texas.
Murphy Boot & Shoe Co., Natick, Mass.
Murray, Elmer, Haverhill, Mass.
Murray, J. H. & Co., Haverhill Mass.
Murray Shoe Co., Ltd., London, Ont., Can.
Murray Shoe Co., Lynn, Mass.
Mutual Shoemakers, Inc., Norridgewock, Me., and No. Anson,
    Me.

N.

Nahm Brothers, Philadelphia, Pa.
Nason & Smith, Haverhill, Mass.
National Shoemakers, Auburn and Lewiston, Me.
Neenah Shoe Co., Neenah, Wis.
Nekervis, Wm. & Co., Philadelphia, Pa.
Nesmith Shoe Co., Brockton, Mass.
Nettleton, A. E., Syracuse, N. Y.
Newcomb-Anderson Shoe Co., Rochester, N. Y.
Newfoundland Boot & Shoe Mfg. Co., Ltd., St. John, New-
    foundland.
Newhall, B. H., Lynn, Mass.
Newman, A. & Co., Philadelphia, Pa.

New Oxford Shoe Co., New Oxford, Pa.
Newton Shoe Co., Newton, N. H.
Newton, J. R. & Co., Philadelphia, Pa.
New York Baby Shoe Co., Brooklyn, N. Y.
Nichols, S. H., Berwick, Nova Scotia.
Nolan-Earl Shoe Co., Petaluma and San Francisco, Cal.
Norman & Bennett, Boston, Mass.
Northern Shoe Co., Duluth, Minn.
North Lebanon Shoe Factory, Lebanon, Pa.
North Star Shoe Co., Minneapolis, Minn.
North Shore Shoe Co., Salem, Mass.
Northwestern Felt Shoe Co., Webster City, Ia.
Noyes-Norman Shoe Co., St. Joseph, Mo.
Noyes, A. B. & Co. (Corp), Georgetown, Mass.
Nu Baby Shoe Co., Lynn, Mass.
Nugent Bros., Beverly, Mass.
Nursery Shoe Co., St. Thomas, Ont., Can.
Nutt, W. H. Shoe Co., E. Boston, Mass.
Nyack Shoe Co-operative, Nyack, N. Y.

O.

O. B. Shoe Co., Drummondsville, Quebec.
Obear, J. L., Lynn, Mass.
Oberholtzer, The G. V. Co., Ltd., Berlin, Ont., Can.
O'Brien, J. F. Shoe Co., Rochester, N. Y.
O'Connor, W. H., Rochester, N. Y.
O'Donnell, J. M. & Co., Brockton, Mass.
O'Donnell Shoe Co., St. Paul, Minn.
Ohio Shoe Co., Lancaster, O.
Ohio State Reformatory, Mansfield, O.
Oliver, Isaac E., Lynn, Mass.
Ordway, A. A. Co., Haverhill, Mass.
Orne, W. P. & Co., Reading, Mass.
Orr, J. K. Shoe Co., Atlanta, Ga.
Osborne, John W., Marbelhead, N. J.
Osgood, C. F. & Co., Hammonton, N. J.

P.

Packard, H. M. & Co., Lynn, Mass.
Packard, L. H. & Co., Montreal, Que., Can.
Packard, M. A. Co., Brockton, Mass.
Packard, Marston & Brooks, Danvers, Mass.
Paff Shoe Mfg. Co., Alexandria, Va.
Paine Shoe Co., Marblehead, Mass.
Palma Shoe Co., Waupun, Wis.
Palmer, The John Co., Ltd., Frederickton, New Brunswick, Can.
Parker Boot & Shoe Mfg. Co., Jefferson City, Mo.

Parker, F. A. & Co., Marblehead, Mass.
Parker, John H., Maiden, Mass.
Parker & Honroe, St. Johns, Newfoundland.
Parsons, James & Co., Brooklyn, N. Y.
Passant & Olley, Philadelphia, Pa.
Peerless Shoe Co., Rochester, N. Y.
Penn Shoe Mfg. Co., Reading, Pa.
Penniman Bros., Middleboro, Mass.
Perfect Shoe Mfg. Co., Brooklyn, N. Y.
Perfection Shoe Co., Rochester, N. Y.
Perry & Malcom Co., Haverhill, Mass.
Pesl, Frank & Co., New York, N. Y.
Peters Shoe Co., St. Louis, Hermann and De Soto, Mo.
Pfeiffer, W. F. & Co., Natick, Mass.
Phelan's, Jeremiah Sons, Rochester, N. Y.
Phelan, James & Sons, Lynn, Mass.
Phillips, Milton, Philadelphia, Pa.
Piehler Shoe Co., Rochester, N. Y.
Piekenbrock, E. B. Shoe Mfg. Co., Dubuque, Ia.
Pierce, S. L. & Co., Cleveland, O.
Pilling, John Shoe Co., Lowell, Mass.
Pincus & Tobias, New York.
Pingree Shoe Co., Detroit, Mich.
Pinkham, L. N., Lynn, Mass.
Pinkham, H. E. Shoe Co., Lynn, Mass.
Pitman, A. A. & Co., Lynn, Mass.
Pittsfield Shoe Co., Pittsfield, N. H.
Plant, Thomas G. Co., Boston, Mass.
Plaut-Butler Shoe Co., Cincinnati, Ohio.
Plymouth B. & S. Co., Plymouth, Pa.
Pohquin & Gagnon, Maisonneuve, Quebec, Can.
Pollok & Halfern, New York.
Pontiac Shoe Mfg. Co., Pontiac, Ill.
Porter, William & Son (Inc.), Lynn, Mass.
Portland Shoe Mfg. Co., Portland, Me.
Portsmouth Shoe Co., Portsmouth, O.
Posner, A., Brooklyn, N. Y.
Pourier Shoe Co., Quebec, Que., Can.
Pratt-Reid Shoe Co., Natick, Mass.
Pray, James A., Weymouth, Mass.
Prenzel, A. H. & Co., Halifax and Dalmatia, Pa.
Priesmeyer, A. Shoe Co., Jefferson City, Mo.
Price, C. H., Burlington, N. J.
Prouty, Isaac & Co., Spencer, Mass.
Purington, J. W., Kingston, N. H.
Puritan Shoe Mfg. Co., Topsfield, Mass.
Putnam, A. H. & Co., Danvers, Mass.
Putnam, H. J. & Co., Minneapolis, Minn.

Putnam & Cross, Lynn, Mass.

Q.

Quaker City Shoe Co., Philadelphia, Pa.
Quakertown Shoe Co., Quakertown, Pa.
Quarryville Shoe Co., Quarryville, Pa.
Quast Shoe Mfg. Co., Louisville, Ky.
Queen Elizabeth Shoe Co., Rochester, N. Y.

R.

Racine Shoe Mfg. Co., Racine, Wis.
Radcliffe Shoe Co., Norway, Me.
Ramsfelder, Erlick Co., Cincinnati, O.
Randall, Adams Co., Lynn, Mass.
Red Fox Shoe Co., Nashville, Tenn.
Red Star Shoe Co., Brooklyn, N. Y.
Red Wing Shoe Co., Red Wing, Minn.
Reed, E. P. & Co., Rochester, N. Y.
Reed, H. B. & Co., Manchester, N. H.
Regal Shoe Co., Whitman, Mass., Milford, Mass.
Regina Shoe Co., Ltd., Montreal, Que., Can.
Rehr Shoe Co., Orwigsboro, Pa.
Reid, Peter, Brooklyn, N. Y.
Reis & Newman, New York, N. Y.
Reliable Shoe Co., Orwigsburg, Pa.
Relindo Shoe Co., Ltd., Toronto, Ont., Can.
Reider-Fisher Shoe Co., Kutztown, Pa.
Reimer, A. H. Shoe Co., Milwaukee, Wis.
Rendell Shoe Co., Trenton, N. J.
Reynolds, R. F., Brockton, Mass.
Reynolds, L. W. Brockton, Mass.
Reynolds, Drake & Gabell, North Easton, Mass.
Rex Shoe Factory, New Orleans, La.
Rice & Hutchins, S. Braintree, Rockland, Marlboro, Mass., and
    Warren, Me.
Rice & Hulbert Co., Cortland, N. Y.
Rich Shoe Co., Milwaukee, Wis.
Richards & Brennan Co., Randolph, Mass.
Richardson, C. H., Lynn, Mass.
Richardson, E. B., Reading, Mass.
Richardson, B. F., Dubuque, Iowa.
Rickard-Gregory Shoe Co., Lynn, Mass.
Rideau Shoe Co., Maisonneuve, Quebec, Can.
Ridgway, A. & Son, Delanco, N. J.
Rindge, Kalmbach, Logie & Co., Ltd., Grand Rapids, Mich.
Ritche, The John Co., Ltd., Quebec, Que., Can.
Riverside Shoe Co., Montreal, Que.

Roberts, Johnson & Rand Shoe Co., St. Louis, St. Charles,
    Hannibal and Washington, Mo., and Jerseyville, Ill.
Rochester Specialty Shoe Co., Rochester, N. Y.
Rochester Soulietta Mfg. Co., Rochester, N. Y.
Rouchette, J. Marcel, Quebec, Que., Can.
Rock Shoe Mfg. Co., Ltd., Quebec, Que., Can.
Rockford Shoe Mfg. Co., Rockford, Ill.
Rogers, A. W. & Son, Raynham, Mass.
Rohrer & Co., Orwigsburg, Pa.
Roland, A. B., Montreal, Que., Can.
Roney & Berger Co., Allentown, Pa.
Roth, Abraham, Baltimore, Md.
Roth Shoe Mfg. Co., Cincinnati, O.
Routier, L., Quebec, Que., Can.
Rowe & Tilton, Haverhill, Mass.
Rowen, Ogg Co., Guelph, Ont., Can.
Royal Shoe Co., Randolph, Mass.
Ruppel, H. & Sons, Brooklyn, N. Y.
Russ, T. M., Salem, N. H.
Russell, W. C. Moccasin Co., Berlin, Wis.

S.

Saalfrank, John, New York, N. Y.
Sachs Shoe Mfg. Co., Cincinnati, O.
Salem Shoe Mfg. Co., Salem, Mass.
Saucony Shoe Mfg. Co., Kutztown, Pa.
Saunders Shoe Co., Camden, N. J.
Sawyer Boot & Shoe Co., Bangor, Me.
Sayre, G. M., Horseheads, N. Y.
Scarth Bros., New York, N. Y.
Scheiffele Shoe Mfg. Co., Cincinnati, O.
Schneider Bros., Natick, Mass.
Schneider, P. C., Allentown, Pa.
Schoenecker, V. Boot & Shoe Co., Milwaukee, Wis.
Schreier, Jos. A., Rochester, N. Y.
Schreijack, John, Brooklyn, N. Y.
Schroeder Shoe Co., Seattle, Wash.
Schryburt, F. & Co., Quebec, Que., Can.
Schultz-Ruck-Delfs Shoe Co., The, Cleveland, O.
Schwartz, I., Brooklyn, N. Y.
Scott, R. H., Marblehead, Mass.
Seattle Shoe Co., Seattle, Wash.
Selby Shoe Co., The Portsmouth, O.
Sellner, Anton, New York.
Seymour & Jackson, Lynn, Mass.
Selz, Schwab & Co., Chicago, Elgin and Genoa, Ill.
Shaft-Pierce Shoe Co., Faribault, Minn.
Sharood Shoe Corp., St. Paul, Minn.
Shaw, A. W. Corp., Freeport, Me.

Shaw, Samuel, Newark, N. J.
Sheboygan Shoe Co., Sheboygan, Wis.
Sheridan Brothers, Haverhill, Mass.
Sherry Shoe Co., Lynn, Mass.
Sheppard & Myer Co., Hanover, Pa.
Sherwood Shoe Co., Rochester, N. Y.
Shick, William, Ferndale, Pa.
Shinboun, P. & Son, Croghan, N. Y.
Shortell, M. & Son, Salem, Mass.
·Simonds & Webster, Haverhill, Mass.
Simons, G. F. & Co., Haverhill, Mass.
Simons, George & Son, N. Weare, N. H.
Sing Sing Prison, Ossining, N. Y.
Skyrm, M. & Co., Cleveland, O.
Slater, Geo. A., Maisonneuve, Quebec, Can.
Slater Shoe Co., Montreal, Que., Can.
Slater & Morrill Shoe Co., So. Braintree, Mass.
Smallwood, F., St. Johns, Newfoundland.
Smaltz, Goodwin Co., Philadelphia, Pa.
Smardon & Percival, Montreal, Que., Can.
Smart, Harris A. & Co., Haverhill, Mass.
Smith, A. F. Co., Lynn, Mass.
Smith-Brisco Shoe Co., Lynchburg, Va.
Smith, F. J. & Co., Haverhill, Mass.
Smith, G. Edwin Shoe Co., Columbus, O.
Smith, McNault Co., Amesbury, Mass.
Smith & Herrick Co., Albany, N. Y.
Smith, J. G. & H. H., Camden, N. J.
Smith, Jeremiah J., Brooklyn, N. Y.
Smith, J. P. Shoe Co., Chicago, Ill.
Smith, M. A. & Son, Philadelphia, Pa.
Snecdicor & Hathaway Co., Detroit, Mich.
Snow, George G. Co., Brockton, Mass.
Solid Leather Shoe Co., Toronto, Ont., Can.
Solomon, J., Brooklyn, N. Y.
Somerset Mfg. Co., Somerset, Pa.
South, E. H. Shoe Co., Bethel, O.
Sovereign Shoe Co., Toronto, Ont., Can.
Spalding, A. G. & Bros. Mfg. Co., Brookyyn, N. Y.
Spaulding, L. H. & Co., Lowell, Mass.
Spieglman & Gluckman, Brooklyn, N. Y.
Spinks, A. B. Mfg. Co., Chicago, Ill.
Spinney, B. F. & Co., Norway, Me. (Radcliffe Shoe Co.)
Sprague Shoe Co., Mechanic Falls, Me.
Springvale Shoe Works, Springvale, Me.
Stacy, Adams & Co., Brockton, Mass.
Standard Mfg. Co., Ltd., Sacksville, New Brunswick, Can.
Standard Shoe Co., Buffalo, N. Y.

Stanley Shoe Co., Boston.
Stanwear Shoe Co., Lynn, Mass.
Star Shoe Co., Davenport, Ia.
Star Shoe Co., Davenport, Ia.
Star Shoe Co., Montreal, Que., Can.
Starner, Copeland Co., Columbus, O., and Marion, O.
Sterner, Henry S., Selins Grove, Pa.
Steel Shoe Co., Racine, Wis.
Sterling Shoe Co., Belleville, Ill.
Sterling Shoe Co., Orangeville, Pa.
Sterling Bros., Ltd., London, Ont., Can.
Stern, Auer & Co., Cincinnati, O.
Stern, Henry, Philadelphia, Pa.
Stetson Shoe Co., S. Weymouth, Mass.
Stevens, J. G., Marblehead, Mass.
Stiles, L. A. & Co., Brooklyn, N. Y.
Stillman, Armstrong Co., Vanceboro, Me.
Stobo, J. M., Quebec, Que., Can.
Stilson-Kellogg Shoe Co., Tacoma, Wash.
Stork Co., Boston, Mass.
Stoneback, C. H., Coopersburg, Pa.
Stoughton Shoe Co., Stoughton, Wis.
Stover & Bean, Lowell, Mass.
Straw, Luther G. Shoe Co., Salem, Mass.
Strickler & Brehm Shoe Co., Hummelstown, Pa.
Strohbeck, Chas. W., Brooklyn, N. Y.
Strong, Geo. Co., East Weymouth, Mass.
Strong & Co., Milwaukee, Wis.
Strootman, John Shoe Co., Buffalo, N. Y.
Strout, Stritter Co., Lynn, Mass.
Sullivan, E. E., Haverhill, Mass.
Sullivan, P. & Co., Cincinnati, O.
Supple Shoe Co., Haverhill, Mass.
Sweet & Savory, Marblehead, Mass.
Sweet, F. F., Haverhill, Mass.
Syracuse Shoe Mfg. Co., Syracuse, N. Y.

T.

Tappan, A. L. Co., Haverhill, Mass.
Tappan Shoe Mfg. Co., Coldwater, Mich.
Taylor, E. E. & Co., Brockton and New Bedford, Mass.
Taylor, H. Shoe Co., Philadelphia, Pa.
Taylor, The Robert Co., Ltd., Halifax, Nova Scotia.
Tebbutt Shoe & Leather Co., Ltd., Three Rivers, Que., Can.
Tennessee Shoe Mfg. Co., Nashville, Tenn.
Tetrault Shoe Co., Montreal, Que., Can.
Thatcher & Co., Richmond, Va.
Thayer, N. B. & Co., East Rochester, N. H.
Thayer, Maguire & Field, Haverhill, Mass.

Thayer & Osborne Shoe Co., Farmington, N. H.
Thivierge, Eugene, Quebec, Que., Can.
Thomas & Co., Brooklyn, N. Y.
Thomas, J. B. & Tarr, Lynn, Mass.
Thompson Bros., Campello, Mass.
Thompson, F. J., Haverhill, Mass.
Thompson & Crocker Shoe Co., Lynn, Mass.
Thompson, Mary J., Rochester, N. Y.
Three K. Shoe Co., The, Milford, Mass.
Thurrel, Batchelder & Co., Calais, Me.
Tie Co., Unadilla, N. Y.
Tilt, J. E. Shoe Co., Chicago, Ill.
Timson & Co., Lynn, Mass.
Todd, Fred S. Co., Rochester, N. Y.
Tourigny & Marols, Quebec, Que., Can.
Travers-Smith Shoe Co., Peabody, Mass.
Trevett & Berry. Lynn, Mass.
Trimble Bros. Co., Calais, Me.
Troy Shoe Co., Troy, Ill.
Tufts & Friedman, Lynn, Mass.
Turner, H. B. & Co., Chillicothe, O.
Twig Shoe Co., Sheboygan, Wis.

U.

Underhill & Sisman Shoe Mfg. Co., Aurora, Ont., Can.
Union Shoe Mfg. Co. (The), Chillicothe, O.
Union Shoe & Slipper Mfg. Co., Chicago, Ill.
Union Shoe Works, Rockford, Ill.
United Slipper Mfg. Co., Chelsea, Mass.
United States Slipper Co., Philadelphia, ·Pa.
United Workingmen's Boot & Shoe Co., San Francisco, Cal.
Upham Bros. Co., Stoughton, Mass.
Utz & Dunn, Rochester, N. Y.

V.

Valentine & Fischer, Waterloo, Ont., Can.
Van Orden, H. C., Rochester, N. Y.
Venor Shoe Co., Rochester, N. Y.
Venor & Co., Buffalo, N. Y.
Vickery, W. P. & Co., Marblehead, Mass.
Victor Shoe Co., Salem, Mass.
Victoria Shoe Co., Ltd. (The), Toronto, Ont., Can.
Vine, H. N., Haverhill, Mass.
Vogel Bros. Shoe Co., Louisville, Ky.
Volquardson, H. F., Davenport, Iowa.

W.

Wagner Bros. Shoe Co., Penn Yan, N. Y.
Walker, J. L. & Co., Lynn, Mass.

Walker-Parker Co., Ltd., Toronto, Ont., Can.
Walkin Shoe Co., Schuylkill Haven, Pa.
Wallace, E. G. & E., Rochester, N. Y.
Walmer, Dan, Huntington, Ind.
Walton, A. G. Co., Chelsea, Mass.
Walton Shoe Co., Wakefield, Mass.
Washington Shoe Mfg. Co. (The), Seattle, Wash.
Waterbury, S. & Son, Brooklyn, N. Y.
Waters, E. H. Co., Harrisburg, Pa.
Watson-Plummer Shoe Co., Dixon, Ill.
Watson Shoe Co., Lynn, Mass.
Watsontown Boot & Shoe Co., Watsontown, Pa.
Wayne Shoe Co., Fort Wayne, Ind.
Weaver, C. B., Ferndale, Pa.
Webb Shoe Mfg. Co., Denver, Colo.
Weber Bros. Shoe Co., North Adams, Mass.
Webster, George L., Haverhill, Mass.
Webster, Ira J., Haverhill, Mass.
Webster, Webber Shoe Co., Haverhill, Mass.
Weil, S. & Co., Brooklyn, N. Y.
Weimer, Wright & Watkin, Philadelphia, Pa. (Lenox Shoe Co.)
Weinbrenner, A. H. Co., Milwaukee, Wis.
Weis, Wm. A. & Co., Wilkesbarre, Pa.
Weisinger, Fred G., Aurora, Ill.
Welch & Kelly, Camden, N. J.
Wells, M. D. Co., DeKalb, Ill.; Fond du Lac and Watertown, Wis., and Chicago, Ill.
Wentworth-Swett Co., Haverhill, Mass.
Werner Shoe Co., Orwigsburg, Pa.
Wertheimer-Swarts Shoe Co., St. Louis, Mo.
Wesson, J. E. & W. G., Worcester, Mass.
Western Shoe Co., Ltd., Berlin, Ont., Can.
Western Shoe Co., Stillwater and St. Paul, Minn.
Western Shoe Co., Janesville, Wis.
Weston Shoe Co., Ltd., Campbellford, Ont., Can.
Weyenberg Shoe Mfg. Co., Milwaukee, Wis.
White, L. Q. Shoe Co., Bridgewater, Mass.
White, Richard, Philadelphia, Pa.
Whitman & Keith Co., Campello, Mass.
Wichert & Gardiner, Brooklyn, N. Y.
Wildomisky, Wm., Brooklyn, N. Y.
Williams, Clark & Co., Lynn, Mass.
Williams, E. W., Winona, Minn.
Williams, Hoyt & Co., Rochester, N. Y.
Williams, Kneeland Co., South Braintree, Mass.
Williams Shoe Co., Brampton, Ont., Can.
Williams Shoe Co., Cincinnati, O.
Willis, F. E. & Co., Lynn, Mass.

Williams, Arthur A. Shoe Co., Holliston, Mass.
Wilson, Charles E., Lynn, Mass.
Wilson, H. C., Toronto, Ont., Can.
Winchell, J. H. & Co., Haverhill, Mass.
Winchester Shoe Co., Buffalo, N. Y.
Winn & Co., Milton, Ont., Can.
Winona Shoe Mfg. Co., Winona, Minn.
Wisconsin Shoe Co., Milwaukee, Wis.
Wise, Shaw &·Feder Co., Cincinnati, O.
Wise & Cooper, Auburn, Me.
Witchell-Sheill Company, Detroit, Mich.
Witham, A. C. & Co., Haverhill, Mass.
Witherell & Dobbins, Haverhill, Mass.
Witt, Geo. D. Shoe Co., Lynchburg, Va.
Wolbold, John, Philadelphia, Pa.
Wolf Bros. & Co., Cincinnati, O.
Wolfe Bros. Shoe Co., Columbus, O.
Wolfe Shoe Mfg. Co., Allentown, Pa.
Wood, G. A. & Co., Beverly, Mass.
Wood, J. T. Co., Ware, Mass.
Wood, R. T. Shoe Co., Burlington, N. J.
Wood & Johnson Co., Rochester, N. Y.
Woodbury, E. S. Co., Salem, Mass.
Woodbury Shoe Co., Beverly, Mass., and Derry, N. H.
Worcester Slipper Co., Worcester, Mass.
Worthley, Mark J., Lynn, Mass.
Wrensch & Herrmann Shoe Co., Milwaukee, Wis.
Wright Co., Berlin, Wis.
Wright Bros., Marblehead, Mass.
Wright, E. T. & Co., Rockland, Mass.
Wright, Peters & Co., Rochester, N. Y.
Wright, Wm., Rochester, N. Y.
Wright & McAdams, Camden, N. J.
Wusterhauser, W. O., Holmes, Pa.
Wyandott Mfg. Co., Kansas City, Mo.

X.

Xenia Shoe Mfg. Co., Xenia, Ohio.

Y.

Yost, C. E. & Bro., York, Pa.
Young, C. L. & Co., Haverhill, Mass.
York Shoe Mfg. Co., York, Pa.
Yuhoke, T. A., Buffalo, N. Y.

Z.

Ziegler Bros., Philadelphia, Pa.
Z. C. M. L., Salt Lake City, Utah.
Zulick, J. S. & Co., Orwigsburg, Pa.

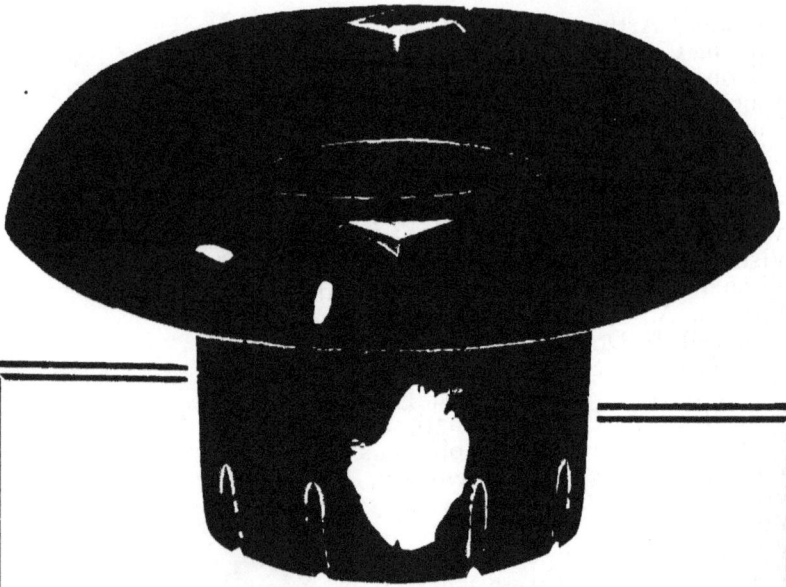

CAN'T WEAR "BRASSY"

You will never have to apologize for a Fast Color Eyelet. Made as they are with tops of solid celluloid they simply can't wear "Brassy" and they do not change color. They preserve their bright new appearance throughout the wear of the shoes. There is only one genuine Fast Color Eyelet and if you will

LOOK FOR THE DIAMOND

trade-mark which is shown in the top of the eyelet illustrated above and insist that it shall appear on all the eyelets with which the shoes you purchase are fitted, you can be sure of getting the genuine. No Fast Color Eyelets are now made which do not bear this little diamond ◆ trademark. It is placed there so that you and your customers can easily recognize the genuine Fast Color Eyelets and avoid imitations.

United Fast Color Eyelet Co. 205 Lincoln St. BOSTON

List of Shoe Wholesalers.

Exclusive Jobbers and the Manufacturers Who Job a Part of Their Shoes.

A.

Adams, Geo. B. Shoe Co., Sioux City, Ia.
Adams, Mears & Co., Boston, Mass.
Adams & Ford, Detroit, Mich., and Cleveland, O.
Aishberg, Edwin, Hartford, Conn.
Ainsworth Shoe Co., Toledo, O.
Albright, H. M., Reading, Pa.
Alexander, Moses, Boise City, Idaho.
Allen Shoe Co., New York, N. Y.
Allen-Bartlett Shoe Co., Burlington, Vt.
American Hand Sewed Shoe Co., Omaha, Neb.
American Shoe & Hat Co., San Antonio, Texas.
Amsterdam Rubber Co., New York, N. Y.
Anthony, Edw. T., Philadelphia, Pa.
Armitage, J., Danvers, Mass.
Arnold, Henegar, Doyle & Co., Knoxville, Tenn.
Arnold, J. M. Shoe Co., Bangor, Me.
Assay & Bretz, Philadelphia, Pa.
Atkinson Guess Co., Paris, Tex.
Atlas Shoe Co., Boston, Mass.
Austin Shoe Co., Wilkesbarre, Pa.

B.

Backhaus, C. A. & Son, New Orleans, La.
Badger State Shoe Co., Milwaukee, Wis.
Baer & Bro., Vicksburg, Miss.
Baker, S. & Son, Kingston, N. Y.
Baker-Holmes Shoe Co., Philadelphia, Pa.
Baker, Poston & Co., Weatherford, Tex.
Baldwin, McGraw & Co., Detroit, Mich.
Baltimore Bargain House, Baltimore, Md.
Baltimore-Harrisburg Shoe Co., Baltimore, Md.
Baltimore Shoe House, Baltimore, Md.
Banner Rubber Co., St. Louis, Mo.
Banner Shoe Co., St. Louis, Mo.
Barnet Shoe Co., Philadelphia, Pa.
Baron's Shoe Store, Portland, Ore.
Barrows, H. D., New London and Norwich, Conn.

Barton Bros., Kansas City, Mo.
Bass & Heard, Rome, Ga.
Bass, A., New York, N. Y.
Batchelder & Lincoln Co., Providence, R. I.
Bates, A. J. & Co., New York City.
Bates, J. E. & Co., New York, N. Y.
Battreall Shoe Co., St. Joseph, Mo.
Bay Shoe Co., Boston, Mass.
Bay State Shoe & Leather Co., New York, N. Y.
Beach, C. R. & Co., Boston, Mass.   •
Beals & Torrey Shoe Co., Milwaukee, Wis.
Bear, M. & Bros., Wilmington, N. C.
Beasley Shoe Co., Lynchburg, Va.
Bedford-Pitman Co., Philadelphia, Pa.
Bell, Walt & Co., Philadelphia, Pa.
Bennett Shoe Co., Boston, Mass.
Bent, Chas. T. A., Boston, Mass.
Bentley & Olmstead Co., Denver, Colo., Des Moines, Ia.
Berg, Chas., Carlisle, Pa.
Berger, R. M. & Co., Peoria, Ill.
Berkshire Shoe Co., Pittsfield, Mass.
Berlow Shoe Co., New York, N. Y.
Berry, A. H. Shoe Co., Portland, Me.
Berryhill-Suther-Durfee Co., Charlotte, N. C.
Betterton & England Shoe Co., Chattanooga, Tenn.
Binghamton Shoe & Rubber Co., Binghamton, N. Y.
Blake & Wheeler Shoe Co., Portland, Me.
Block & Kohner Merc. Co., St. Louis, Mo.
Bode-Larson Shoe Co., Keokuk, Ia.
Bornstein, M. & Son, Boston, Mass.
Boston Shoe Co., New York, N. Y.
Bowne-Gaus Shoe Co., Utica, N. Y.
Boyd, Thomas & Co., New York, N. Y.
Bradley & Metcalf Co., Milwaukee, Wis.
Brand Shoe Co., Roanoke, Va.
Brandt, H. & Sons, Chicago, Ill.
Brauer & Leventhal, New York, N. Y.
Branch Bros., Lincoln, Neb.
Brien Bros., Boston, Mass.
Britt-Carson Shoe Co., Columbus, Ga.
Brown-Evans Co., Charleston, S. C.
Brown & Ross Co., Knoxville, Tenn.
Brown Shoe Co., St. Louis, Mo.
Brunel, D. W. Shoe Co., Portland, Me.
Buckingham & Hecht, San Francisco, Cal.
Bultman Bros., Sumter, S. C.
Burke & Richards Co., Boston, Mass.
Burlington Rubber Shoe Co., Burlington, Ia.

Burney, A. S., Rome, Ga.
Burrow, Jones & Dyer Co., St. Louis, Mo.
Burt & Packard Co., Boston, Mass.
Butler Bros., St. Louis, Mo.
Butler & Tyler Co., New Haven, Conn.

C.

Cable, H. W. Guillford, N. Y.
Cady-Ivison Shoe Co., Cleveland, O.
Cahn, Nickelsburg & Co., San Francisco, Cal.
Carlatt-Stetzler-Hatler Shoe Co., Kansas City, Mo.
Carroll Adams & Co., Baltimore, Md.
Carruthers-Jones Shoe Co., Memphis, Tenn.
Carruthers-Jones Shoe Co., St. Louis, Mo.
Catlin & Knox, Leavenworth, Kan.
Central Shoe Co., Indianapolis, Ind.
Central Shoe & Rubber Co., Syracuse, N. Y.
Central Shoe Mfg. Co., Philadelphia, Pa.
Chaddock, W. H., Pittsburg, Pa.
Chapin, Geo. A., Belleville, Kan.
Chestnut & Barentine, Wilmington, N. C.
Childs, H. & Co., Pittsburg, Pa.
Chipman, Harwood & Co., Boston, Mass.
Claflin, Thayer & Co., New York, N. Y.
Clark-Hutchinson Co., Boston, Mass., New York, N. Y., Pittsburg, Pa., and Providence, R. I.
Clauss Bros., Allentown Pa.
Cohen-Adler Shoe Co., Baltimore, Md.
Cohen, Dan Co., Cincinnati, O.
Cohen, I., Boston, Mass.
Cohen, M. & Sons, Lebanon, Pa.
Cohen, S., Boston, Mass.
Cohen Shoe Co., Milwaukee, Wis.
Cohn, Goldwater & Co., Los Angeles, Cal.
Collins, J. D., Spartanburg, S. C.
Columbia Jobbing House, Scranton, Pa.
Congress Shoe & Rubber Co., Boston, Mass.
Continent Shoe Co., Chicago, Ill.
Converse, F. & Son, Troy, N. Y.
Cosby Shoe Co., Lynchburg, Va.
Cosgrave, W. B. Shoe Co., Zanesville, Ohio.
Cox, A. F. & Son, Portland, Me.
Cox, C. P., Rochester, N. Y.
Craddock-Terry Co., Lynchburg, Va., Baltimore, Md.
Cross Shoe Co., Denver, Colo.
Crowder-Cooper Shoe Co., Indianapolis, Ind.
Culter-Seip Co., Chillicothe, O.
Cumberland County Shoe Co., Wheeling, W. Va.

Currier, H. M. & Co., Boston. Mass.
Currier, C. E. Shoe Co., Boston, Mass.
Currin, W. G., Baltimore, Md.
Curtis & Co., Marlin, Texas.
Curtis, N. & Co., Boston, Mass.
Cushman, W. C. & Co., Boston, Mass.
Cutler & Porter, Springfield, Mass.

D.

Dallas Bros., Philadelphia, Pa.
Daniels, George F. & Co., Boston, Mass.
Dannenberg Co., Macon, Ga.
Dasenbrook-Sales Co., Chicago, Ill.
Day Rubber Co., St. Louis, Mo.
DeCou Bros., Co., Philadelphia, Pa.
Des Moines Rubber Co., Des Moines, Ia.
Detroit Rubber Co., Detroit, Mich.
Detroit Wholesale Auction Co., Boston, Mass.
Diamond Shoe Co., New York, N. Y.
Diamondstone, L., Pittsburg, 'Pa.
Dils, H. P. & J. W., Parkersburg, W. Ca.
Dinsmore & Son, B. C., Belfast, Me.
Dittmann, Geo. F. Boot and Shoe Co., St. Louis, Mo.
Dixon-Bartlett Co., Baltimore, Md.
Dobrein, M., Boston, Mass.
Dolge, Alfred Felt Co., New York, N. Y.
Don Shoe Co., Boston, Mass.
Donat, A. I. & Co., Chicago, Ill.
Dorell, W. H. & Son, Philadelphia, Pa.
Doty-Bennett Shoe Co., Chicago, Ill.
Dougherty-Fithian Shoe Co., Portland, Ore.
Dovenmuehle, H. F. C. & Son, Chicago, Ill.
Down & Tryon, Philadelphia, Pa.
Drake, Innes, Green Shoe Co., Charleston, S. C.
Dunham Bros., Providence, R. I., & Brattleboro, Vt.
Dunlap, D. R. Mercantile Co., Mobile, Ala.
Dunn, J. P. Shoe & Leather Co., Denver, Colo.
Dunn-Salmon Co., Syracuse, N. Y.
Durrell Bros., Cincinnati, O.

E.

Eastern Shoe Mfg. Co. (Essex Shoe Co.), Boston, Mass.
Edwards Shoe Co., Boston, Mass.
Egan, E. J. & Co., San Francisco, Cal.
Eichengreen & Co., Baltimore, Md.
Eisenman, J., Pittsburg, Pa.
Ellett, Kendall Shoe Co., Kansas City, Mo.

Elmira Shoe Co., Philadelphia, Pa.
Emery, J. W., Boston, Mass.
Emery & Co., LJacksonville, Fla.
Empire Specialty Co., Boston, Mass.
Empire State Rubber Co., Amsterdam, N. Y.
Endicott-Johnson Co., St. Louis, Mo., Pittsburg, Pa., and New York, N. Y.
Eveland, C. S., Chicago, Ill.

F.

Falconer, Holt & Co., Boston, Mass.
Faller's, Isaac Sons Co., Cincinnati, O.
Fanger & Rampe, Cincinnati, O.
Fargo, Keith & Co., Chicago, Ill.
Farnham, Geo. W. Co., Buffalo, N. Y.
Farnum, I. P., Chicago, Ill.
Farnsworth, B. B. Shoe Co., Portland, Me.
Farr Bros. & Co., Allentown, Pa.
Faucette- Peavier Shoe Co., Bristol, Tenn.
Filsinger-Boette Shoe Co., St. Louis, Mo.
Finck's, August, Sons, Syracuse, N. Y.
Finkovitch, M., Boston, Mass.
Fisher, Nathaniel & Co., New York, N. Y.
Fleishman, Morris & Co., Richmond, Va.
Flower City Shoe Co., Rochester, N. Y.
Foot, Schulze & Co., St. Paul, Minn.
Forner & Purviance, Pittsburg, Pa.
Forney Bros. Shoe Co., Harrisburg, Pa.
Forsyth, R. & Son, Buffalo, N. Y.
Forwood Shoe Mfg. Co., Cincinnati, O.
Frank, P. E. & Co., Fall River, Mass.
Frank, Wm. Co., Paris, Texas.
Frank & Adled, Baltimore, Md.
Frankie, Louis, Denver, Colo.
Franklin Shoe Co., Boston, Mass.
Frehling, S. & Son, Chicago, Ill.
French, G. R. & Sons, Wilmington, N. C.
Fried, L., New York, N. Y.
Friedman, B., New York City.
Friedman-Shelby Shoe Co., St. Louis, Mo., and Boston, Mass.
Friedman & Cohnreich Co., San Francisco, Cal.
Friendly Boot & Shoe Co., Elmira, N. Y.

G.

Gaines Kennedy Co., Knoxville, Tenn.
Galveston Shoe & Hat Co., Galveston, Tex.
Gardner, L., Boston, Mass.

Gates, John & Co., Cincinnati, O.
Geddes Brown Shoe Co., Indianapolis, Ind.
George & Marvin Shoe Co., San Francisco, Cal.
Gibbon, C. D. & Son, Philadelphia, Pa.
Ginsburg, Louis, Worcester, Mass.
Glasburg, D., Boston, Mass.
Glaser, Jos., Boston, Mass.
Globe Jobbing Co., Worcester, Mass.
Goldberg Bros. & Co., Phoenix, Ariz.
Golden State Shoe Co., Los Angeles, Cal.
Goldman & Epstein, Boston, Mass.
Goldsmith Bros., Scranton, Pa.
Coldstein, Jos., Boston, Mass.
Goldstein, Julius, Boston, Mass.
Goldstein, N. T., Boston, Mass.
Goldstein, S., Boston, Mass.
Goodbar & Co., Memphis, Tenn.
Goodman Bros. Shoe Co., Portland, Ore.
Goodwear Shoe Co., Belleville, Ill.
Goodyear Rubber Co., St. Paul, Minn., Kansas City, St. Louis,
    Mo., and Portland, Ore.
Gordon Shoe Co., Quincy, Ill.
Gotzian, C. & Co., St. Paul, Minn.
Graming-Spalding Co., Atlanta, Ga.
Grand Rapids Shoe & Rubber Co., Grand Rapids, Mich.
Great Northern Rubber Co., Detroit, Mich.
Great Northern Shoe Co., Boston, Mass.
Green & Sons, Louisville, Ky.
Greene, Anthony & Co., Providence, R. I.
Greene, H. N. & Son, Kansas City, Mo.
Grieb, J. G. & Sons, Philadelphia, Pa., Baltimore, Md., Wilkes-
    barre and Reading, Pa.
Griffith, C. D. Shoe Co., Denver, Colo.
Grimsrud Shoe Co., Minneapolis, Minn.
Grossman, A., New York, N. Y.
Grotjan, Lobe Co., Boston, Mass.
Groves & Rood, Chicago, Ill.
Gustin, Edw. E., Lincoln, Neb.
Guthmann, Carpenter & Telling, Chicago, Ill.

H.

Haas, N. A. & Son, Allentown, Pa.
Haas Merc. Co., St. Louis, Mo.
Hahn, F. W. & Co., Rochester, N. Y.
Hahn & Rampe Co., Rochester, N. Y.
Halle's, S. Sons, Baltimore, Md.
Hamburger Bros., St. Louis, Mo.
Hamilton-Brown Shoe Co., St. Louis, Mo., also Boston, Mass.

Hanford, H. B. Co., Philadelphia, Pa.
Hanke, Charles, J., Chicago, Ill.
Hannah, W. D. Shoe Co., New York City.
Hannah, McCarthy Shoe Co., Auburn, N. Y.
Hanover Shoe Co., Boston, Mass.
Harper, Kirschten Shoe Co., Chicago, Ill.
Harris, Wm. & Sons, Philadelphia, Pa.
Harrison Shoe Co., Boston, Mass.
Hartray, J. P. Shoe Co., Chicago, Ill.
Hasbrouck, Gilford, Kingston, N. Y.
Hathaway Shaft Shoe Co., Minneapolis, Minn.
Haupt, J. & Co., Baltimore, Md.
Haverhill Shoe Co., Boston, Mass.
Hay, J. S. Shoe Co., Los Angeles, Cal.
Haynes Henson Shoe Co., Inc., Knoxville, Tenn.
Hayward Bros. Shoe Co., Omaha, Neb.
Hean & Molly Shoe Co., Harrisburg, Pa.
Heilbrunn, J. & Son, Rochester, N. Y.
Henderson-Black Co., Troy, Ala.
Herman Bros., Lindauer & Co., Nashville, Tenn.
Herold-Bertsch Shoe Co., Grand Rapids, Mich.
Hessberg, Magnus, Richmond, Va.
Higman Shoe Co., Sioux City, Ia.
Hill, J. E., Boston, Mass.
Hill, J. F. Shoe Co., Boston, Mass.
Hinkle Shoe Co., Evansville, Ind.
Hirsch, H. & Bro., New York, N. Y.
Hirschberg, S. & Co., Boston, Mass.
Hirth, Krause & Co., Grand Rapids, Mich.
Hixon Shoe Co., Boston, Mass.
Hobbs, F. H., New York, N. Y.
Hollins, Sons & Co., Nashville, Tenn.
Hollister & Noble, Auburn, N. Y.
Homer Shoe Co., Oxford, Mass.
Hosmer, Codding Co., New York & Boston.
Hostettler, W. J. Shoe Co., Erie, Pa.
Hub Shoe Co., Boston, Mass.
Hunt, N. A. & Co., Charleston, S. C.
Hurd & Fitzgerald Shoe Co., Utica, N. Y.
Hutchinson Shoe Co., Jacksonville, Fla.

I.

Ideal Shoe Co., Boston, Mass.
Interstate Shoe Co., Boston, Mass.
Interstate Rubber Co., Omaha, Neb.
Iroquois Rubber Co., Buffalo, N. Y.

J.

Jackson, H. W., Wheeling, W. Va.
Jacobs Bros., New York, N. Y.
Jandorf, R. & Co., Baltimore, Md.
Jantzen Shoe Co., Philadelphia, Pa.
Jellerson, J. L., Boston, Mass.
Jesselman, H. L., Boston, Mass.
Jewett Co., Gloucester, Mass.
Johansen Bros. Shoe Co., San Francisco, Cal., and St. Louis,
   Mo.
Johnson, H. D., Altoona, Pa.
Johnson, Henry E. Co., Chicago, Ill.
Johnson, Moody & Co., Boston, Mass.
Jolles, Chas., Boston, Mass.
Jolles, I., Boston, Mass.
Jolles, Louis, Boston, Mass.
Jones, Alfred, Columbus, O.
Jones, James, Albany, N. Y.
Jones-Kennington, D. G. Co., Jackson, Miss.
Jones, W. R., Tamaqua, Pa.
Jordan, G. B. Co., Oakland, Cal.
Jung Shoe Co., Shelboygan, Wis.

K.

Kanawha Shoe Co., Charleston, W. Va.
Kaplan, M. & Sons, Boston, Mass.
Katz &Israel Shoe Co., Cincinnati, O.
Katzenberg, M. & Sons, New York, N. Y.
Keiffer Bros., New Orleans, La.
Keiffer, Isadore & Co., New Orleans, La.
Kelly, C. J. & Co., Boston, Mass.
Kelly, Wm. H., Providence, R. I.
Kenimer-Irwin Shoe Co., Birmingham, Ala.
Kern, Lauderbach & Co., Philadelphia, Pa.
Kimball, H. H. & Co., Kansas City, Mo.
King Bros. Shoe Co., Bristol, Tenn.
Kiser, M. C. Co., Atlanta, Ga.
Klinordinger, A. W., Pittsburg, Pa.
Knorr & Ruth, Reading, Pa.
Koch Brothers, Chicago, Ill.
Koehn Brothers, Chicago, Ill.
Kohlman, Isaac, New Orleans, La.
Kornman, Sawyer & Co., Nashville, Tenn.
Korick, Chas., Phoenix, Ariz.
Kraeger, O. H., New York, N. Y.
Krausse Bros., Portland, Ore.
Kreider-Cushman Co., Cedar Rapids, Iowa.

Krieder Baker Shoe Co., New York City.
Kuttner, J. & Co., Rome, Ga.

L.

La Bonte, Geo. N. Shoe Co., Boston, Mass.
Laird & Taylor Co., Pittsburg, Pa.
Lamkin & Foster, Boston, Mass.
Lane Bros. Co., Boston, Mass.
Lane Shoe Co., St. Louis, Mo.
Lane, R. H. & Co., Toledo, O.
Lang, H. J. Shoe Co., Pittsburg, Pa.
Langley, J. E., Detroit, Mich.
Lantzky-Allen Shoe Co., Dubuque, Ia.
Lapham, Geo. H., Boston, Mass.
Lapinsky, J., New York City.
Leatherbury Shoe Co., Clarksburg, W. Va.
Leaverton, G. W., Springfield, Ill.
Lebby Shoe Co., Charleston, S. C.
Lebovitz, B., New York, N. Y.
Leh, H. & Co., Allentown, Pa.
Lehigh Shoe & Rubber Co., Allentown, Pa.
Levy, Benjamin & Co., Scranton, Pa.
Levy, Morris, Wilmington, Del.
Levy, Sam & Co., Nashville, Tenn.
Levy, M. & Co., New York, N. Y.
Levy, Wolff & Pitts Shoe Co., Montgomery, Ala.
Lewis, O. J. Merc. Co., St. Louis, Mo., and Boston, Mass.
Lighthiser, T. A., Honesdale, Pa.
Liles, W. H., Wadesboro, N. C.
Lippincott, Son & Co., Boston, Mass.
Lobdell, H. H. Co., Chicago, Ill., & St. Louis, Mo.
Locke Shoe Co., Wheeling, W. Va.
Long & Davidson, Lancaster, Pa.
Love, J. W. & Co., Minneapolis, Minn.
Lovejoy, E. M. & Co., Boston, Mass.
Lowenberg, D. Boot & Shoe Co., Norfolk, Va.
Lynchburg Shoe Co., Lynchburg, Va.
Lynn-Beverly Shoe Co., New York, N. Y.
Lyons, F. C. Shoe Co., New York City.

M.

Maas Shoe Co., Boston, Mass.
Mackey Shoe Co., Sedalia, Mo.
Magnes Shoe Co., San Francisco, Cal.
Mahoney Cox & Co., Bristol, Tenn.
Manchester Shoe Co., Boston, Mass.
Manhattan Shoe Co., New York, N. Y.

Mann & Longini, Cincinnati, O.
Manning Shoe & Rubber Co., Bsoton, Mass.
Marks, C. W. Shoe Co., Chicago, Ill.
Marks, Oscar & Son, Newberne, N. C.
Marks, Rothenberg & Co., Meridian, Miss.
Marks & Stix, Cincinnati, O.
Markwell, L. D. & Co., Milwaukee, Wis.
Marshall Bros.. Milwaukee, Wis.
Martinez, W. J. & Bro., New Orleans, La.
Marvel Shoe Co., Boston, Mass.
Marx, W. S. Shoe Co., Cincinnati, O.
Marx, B. & Son, Detroit, Mich.
Maryland Shoe Co., Cumberland, Md.
Master & Hoffman, Philadelphia, Pa.
Matchless Shoe Co., Boston, Mass.
Mathias, Albert & Co., El Paso, Texas.
Mathis, J. B. Shoe Co., Greensboro, N. C.
Maury, Henry Shoe Co., New York, N. Y.
Mauzy & Reid Shoe Co., San Francisco, Cal.
Mayer, F. Boot & Shoe Co., Seattle, Wash., and Milwaukee,
    Wis.
Mayhew, P. G. Shoe Co., Grand Rapids, Mich.
Mayo, W. F. Co., New York, N. Y., and Boston, Mass.
McCarthy, A. F. & Co., Oswego, N. Y.
McCarthy, R. B. & Co., Boston, Mass.
McComb, F. E. Shoe Co., Scranton, Pa.
McConnell, S. E. & Son, Atlanta, Ga.
McCord-Donovan Shoe Co., St. Joseph, Mo.
McCune, C. A. Co., Des Moines, Ia.
McDonnald, J. I., Martinsburg, W. Va.
McGarry, D. P. Co., Rochester, N. Y.
McIlvaine, J. F. Co., Philadelphia, Pa.
McIntosh Co., Springfield, Mass.
McMillan & Hazen Co., Knoxville, Tenn.
Meaney, Joseph I. & Co., Philadelphia, Pa.
Meis, Charles Shoe Co., Cincinnati, O.
Meltzer, Frank, Jr., New York City.
Meize-Alderton Shoe Co., Saginaw, Mich.
Mercer, Thos. S. & Co., Pittsburg, Pa.
Merchants' Rubber Co., New York, N. Y.
Merritt, Elliott & Co., New York, N. Y.
Metropolitan Shoe Co., New York, N. Y.
Metzler, Julius, New York.
Meyer & Elkan, Selma, Ala.
Meyer, Milton, J. & Co., New York, N. Y.
Michigan Shoe Co., Detroit, Mich.
Midgley, J. W., Kansas City, Mo.
Miles, W. H. Shoe Co., Richmond, Va.

Miller, J. Co., Racine, Wis.
Miller Bros. Co., Chattanooga, Tenn.
Miller Bros., Galveston, Texas.
Miller, Simon & Son, Philadelphia, Pa.
Mills, George T. & Co., Baltimore, Md.
Minowitz F., Leadville, Colo.
Mistrot Bros. & Co., Galveston, Texas.
Monroe Brothers & Co., Philadelphia, Pa.
Montague Rubber Co., Norfolk, Va.
Montana Shoe Co., Butte, Mont.
Montayne & Co., Kingston, N. Y.
Monteleone, A., New Orleans, La.
Montgomery, Ward & Co., Boston, Mass.
Moore, E. G. & Co., Plattsburg, N. Y.
Morgan, K. B., Kansas City, Mo.
Morgan Bros. & Co., Wilkesbarre, Pa.
Morrison, D. J. & Son, Savannah, Ga.
Morrison, John C., Philadelphia, Pa.
Morse & Rogers, New York, N. Y., and Pittsburg, Pa.
Morse, Chas., Boston, Mass.
Moses, K., Boston, Mass.
Moses, S. P., Boston, Mass.
Motteler Shoe Co., Louisville, Ky.
Mullen Shoe Co., Pittsburg, Pa.
Mundt, S. L., Helena, Ark.
Murray, Dibrell Shoe Co., Nashville, Tenn.

N.

Naftel-Nicrosi D. G. Co., Montgomery, Ala.
National Shoe Mfg. Co., Richmond, Va.
Nettleton, A. C. & Son, Scranton, Pa.
Newell Bros., Troy, Pa.
Nolan & Earl Shoe Co., San Francisco & Los Angeles, Cal.
North Star Shoe Co., Minneapolis, Minn.
Northwestern Shoe Co., Seattle, Wash., and Milwaukee, Wis.
Norton Berger Shoe Co., Little Rock, Ark.
Novelty Shoe Co., Boston, Mass.
Noyes, Norman Shoe Co., St. Joseph, Mo.

O.

O'Connor Shoe Co., Chicago, Ill.
Ohio Valley Shoe Co., Cincinnati, O.
Ohms & Jung Shoe Co., Belleville, Ill.
Olcovich, Emil, Las Angeles, Cal.
Olympia Shoe Co., Boston, Mass.
Omaha Rubber Co., Omaha, Neb.
Oppenheim, M., New York, N. Y.

Orr, J. K. Shoe Co., Atlanta, Ga.
Ottenheimer, H. L., Kansas City, Mo.
Outing Shoe Co., Boston, Mass.

P.

Packard, F. M. & Co., Boston, Mass.
Parker, Holmes & Co., Boston, Mass., New York, N. Y., and
 Pittsburg, Pa.
Patapsco Shoe Co., Baltimore, Md.
Patton & Hall, Schenectady, N. Y.
Paul Bros., Philadelphia, Pa.
Payne Shoe Co., Charleston, W. Va.
Pearl Bros., Boston, Mass.
People's Shoe Co., Mobile, Ala.
Perry, Geo. W. & Co., St. Louis, Mo.
Peterman, D. S. & Co., York, Pa.
Peters Shoe Co., St. Louis, Mo.
Peterson Shoe Co., Keokuk, Ia.
Pfeiffer, G. W., Boston, Mass.
Phillips Big Store Co., Ottumwa, Iowa.
Phillips, W. R. & Co., Richmond, Va.
Philips, Wm. H., Providence, R. I.
Piekenbrock, E. B. Shoe Mfg. Co., Dubuque, Ia.
Pilgrim Rubber Co., Boston, Mass.
Pillow Shoe Co., Boston, Mass.
Pincus, Max & Co., Norfolk, Va.
Pingree Bros. Co., Ogden, Utah.
Pinkham, H. E. Shoe Co., Boston, Mass.
Pitcher, R. N. Shoe Co., Milwaukee, Wis.
Plant, Thos. G. Co., St. Louis, Mo.
Plaut, Nathan & Co., Cincinnati, O.
Poole Shoe Co., Philadelphia, Pa.
Porter, J. H., Boston, Mass.
Powell Bros. Shoe Co., Pittsburg, Pa., and New York City.
Powell & Campbell, New York, N. Y.
Pretzfelder, H. & Co., Baltimore, Md.
Prince Shoe Co., Portland, Ore.
Prior, W. A. Co., Boston, Mass.
Pritchard Shoe Co., Pittsfield, Mass.
Putnam Shoe Co., Boston, Mass.
Putnam, A. A. & Son, Chicago, Ill.
Putney, Stephen Shoe Co., Richmond, Va.

Q.

Quirl, E. A., Amsterdam, N. Y.

R.

Radey, Frank H., Philadelphia, Pa.

Raff, Henry D. & Co., Chicago, Ill.
Ray, Thomas K. & Co., Pittsburg, Pa.
Reed, E. P. & Co., San Francisco, Cal.
Regent Shoe Mfg. Co., Lincoln, Neb.
Reynolds, Drake & Gabell, Detroit, Mich.
Rice & Hutchins, Atlanta, Ga.; Baltimore, Md.; St. Louis, Mo.,
   and Cleveland, Ohio.
Richard, A. & Co., New York, N. Y.
Richardson Bros. Shoe Co., Nashville, Tenn.
Richmond, Carl, Boston, Mass.
Rieley, E. A., New Albany, Ind.
Riemer, A. R. & Co., Milwaukee, Wis.
Rindge, Kalmbach, Logie & Co., Grand Rapids, Mich.
Riverside Shoe Co., St. Louis, Mo.
Robbins, N. E., Boston, Mass.
Roanoke Shoe Co., Roanoke, Va.
Roberts & Hoge, Richmond, Va.
Roberts, Johnson & Rand Shoe Co., St. Louis, Mo.
Rochester Rubber Co., Rochester, N. Y.
Rochester Shoe Specialty Co., Rochester, N. Y.
Roenitz, H. C., Oshkosh, Wis.
Rogers, H. M. Co., San Francisco, Cal.
Rogers, Wormer & Co., Cincinnati, Ohio.
Rose, Adolph & Co., Vicksburg, Miss.
Rosenbaum, A., Portsmouth, Va.
Rosenberg, B. & Sons, New Orleans, La.
Rosenberg, S. & Co., Boston, Mass.
Rosenberg, S. Co., Cambridge, Mass.
Rosenbloom Shoe Co., New York City.
Rosenbush, A. A. & Co., Boston, Mass.
Rosenheim, Jos. Shoe Co., Savannah, Ga.
Ross, L. P., Rochester, N. Y.
Rubel, A. & Co., Corinth, Miss.
Rubel, J. & Co., Okolona, Miss.
Ruff, Alf & Son, Butler, Pa.
Runkel, Louis, Cincinnati, O.
Russell, R. J., La Crosse, Wis.

S.

Sacks, A. M. & Co., Boston, Mass.
Samuels, M. & Co., Baltimore, Md.
Sanders Duck & Rubber Co., St. Louis, Mo.
Sandler, M., Boston, Mass.
Sanger Bros., Dallas & Waco, Tex.
Sawyer, P. A. & Co., New York City.
Sawyer Boot & Shoe Co., Bangor, Me.
Schliek Shoe Mfg. Co., St. Paul, Minn.
Shoenecker, V. B. & S. Co., Milwaukee, Wis.

Schoonmaker, L. E. & Co., New York.
Schulten, J. J. & Co., Louisville, Ky.
Schwarz & Sallenbach, St. Louis, Mo.
Scowden & McAllister, Columbus, O.
Scowcroft, J. & Sons Co., Ogden, Utah.
Seehrmann & Cross, Wilkesbarre, Pa.
Seifter, P. Co., New York City.
Selz-Schwab & Co., Pittsburg, Pa.
Semmelman, C. & Sons, West Point, Miss.
Sharood Shoe Corporation, St. Paul, Minn.
Sidwell, DeWindt Shoe Co., St. Louis, Mo., and Chicago, Ill.
Silverman, J., Boston, Mass.
Simmons Boot & Shoe Co., Toledo, O.
Simon, Jacob, Dover, Del.
Simpson, H. L. & Son, Alexandria, Va.
Singer Shoe Co., Boston, Mass.
Sinsheimer Bros. & Co., Chicago, Ill.
Sisson, C. S. Co., Boston, Mass., & Providence, R. I.
Smartt Bros. & Co., Chattanooga, Tenn.
Smith & Herrick Co., Albany, N. Y.
Smith-Briscoe Shoe Co., Lynchburg, Va.
Smith-Coghill Shoe Co., Fredericksburg, Va.
Smith, H. E. & Son, Worcester, Mass.
Smith, J. E., Jr. & Co., Dublin, Ga.
Smith, R. P. & Sons, Chicago, Ill.
Smith, Schaffer & Co., Philadelphia, Pa.
Smith, W. C. & Son, Elberton, Ga.
Smith, Wallace Shoe Co., Chicago, Ill.
Smith, Wm. Sumner, Chicago, Ill.
Snyder Shoe & Rubber Co., Easton, Pa.
Sobel, A., Boston, Mass.
Sosner, S. M., Springfield, Mass.
Spear Bros. Co., Baltimore, Md.
Specialty Shoe Co., Rochester, N. Y.
Speise, F. P. & Co., Tamaqua, Pa.
Standard Boston Shoe Co., Boston, Mass.
Standard Shoe Co., New York, N. Y.
Stanton's, C. A. Sons, Madison, Ind.
Star Shoe Co., Davenport, Ia.
Stein, A., New York, N. Y.
Stella, Joseph, New York, N. Y.
Stern & Co., Richmond, Va.
Sterling Shoe Co., New Haven, Conn.
Stevens, Jas. E., Providence, R. I.
Stevens, L. F., Boston, Mass.
Stevens, S. W. & Co., Chicago, Ill.
Stewart Bros. & Co., Pittsburg, Pa.
Stewart-Dawes Shoe Co., Los Angeles, Cal

St. Paul Rubber Co., St. Paul, Minn.
Stickney, Niles T., Milwaukee, Wis.
Stoebener, Geo., Jr., Pittsburg, Pa.
Streng & Thalheimer, Louisville, Ky.
Superba Shoe Co., Rochester, N. Y.
Swarts, C. L., Boston, Mass.

T.

Tampa Hide & Skin Co., Tampa, Fla.
Tapley, A. P. & Co., Boston, Mass.
Tapp, J. L. Co., Columbia, S. C.
Taylor, B. L., Watertown, N. Y.
Tedcastle, A. W. & Co., Boston, Mass.
Terrell, C. C. & Co., Cairo, Ill.
Thacher & Co., Philadelphia, Pa.
Thing, G. E. & Co., Buffalo, N. Y.
Thing, S. B. & Co., Boston, Mass.
Timson Bros., Boston, Mass.
Thomas, S. J. & Co., Norfolk, Va.
Thomas Shoe Co., Charleston, W. Va.
Thomson-Crooker Shoe Co., Boston, Mass.
Top Round Shoe Co., Boston, Mass.
Toppino, Seidenbach & Larose, New Orleans, La.
Tubman, The Robt. E. Co., Baltimore, Md.
Tucker & Hagen, Chicago, Ill.
Turner-Tomkins Shoe Co., Philadelphia, Pa.
Turrell Bros., Tacoma, Wash.
Tracy Shoe Co., Portsmouth, O.

U.

Upham, Gordon & Co., Springfield, Mo.

V.

Vance Shoe Co., Wheeling, W. Va.
Vinsonhaler Shoe Co., St. Louis, Mo.
Vogel Bros. & Co., Louisville, Ky.

W.

Wagner Bros., Pittsburg, Pa.
Wagner, H. J., Providence, R. I.
Waldron Shoe Co., Boston, Mass.
Walker Bros. D. G. Co., Salt Lake City, Utah.
Walker, Wm. H. & Co., Buffalo, N. Y.
Wallace & Son, York, Pa.
Walters, W. W. Co., Rochester, N. Y.
Warren Rubber Co., Warren, O.

Wayne Shoe Mfg. Co., Fort Wayne, Ind.
Waxelbaum, E. A. & Bro., Macon, Ga.
Weil, E. A. & Co., Savannah, Ga.
Weil, H. & Bro., Coldsboro, N. C.
Weimer, Wright & Watkin, Philadelphia, Pa.
Wells, M. D. Co., Chicago, Ill.
Wells-Curtis Shoe Co., Columbus, Ga.
Werner, Henry C. Co., Columbus, O.
Werner Shoe Co., Zanesville, Ohio.
Wertheimer-Swarts Co., St. Louis, Mo., and Boston, Mass.
West, Geo. H. Shoe Co., Philadelphia, Pa.
Western Shoe Co., St. Paul, Minn., Kansas City, Mo., and
   Toledo, O.
Western Shoe Makers, Louisville, Ky.
Wheeler, F. C., Kansas City, Mo.
White, J. B. & Co., Augusta, Ga.
White Shoe & Rubber Co., Worcester, Mass.
Whitney, Wabel & Co., Cleveland, Ohio.
Whittinghill-Harlow Shoe Co., St. Joseph, Mo.
Williams-Marvin Co., San Francisco, Cal., and Boston, Mass.
Willis, E. L. Co., Boston, Mass.
Wilner, H. D., Green Bay, Wis.
Winch Bros., Boston, Mass., and New York, City.
Wingo, Ellett & Crump Shoe Co., Richmond, Va.
Winn-Jones Shoe Co., Valdosta, Ga.
Winslow, C. R. & Co., Portland, Ore.
Wolfman, Nathan, Boston, Mass.
Wright, Augustus Co., Petersburg, Va.

Y.

Young, Edward P. Co., Boston, Mass.
Young, Geo. H., San Francisco, Cal.
Young, J. R., Springfield, Ill.
Young, W. J. & Co., New York, N. Y.
Young Shoe Co., Selma, Ala.

Z.

Zeman & Collins Shoe Co., Pittsburg, Pa.

Revised Shoe Glossary.

Terms of the Shoe and Leather Trade Explained in a Clear, Concise Manner.

BACKSTAY—A strip of leather covering and strengthening the back seam of a shoe; "English backstay," one that meets the quarters on each side and is sewed to them, forming the lower part of the back of the shoe; "California backstay," a term sometimes applied to piping caught in the back seam.

BAL, (an abbreviation of the word "Balmoral")—A front lace shoe, of medium height, men's, women's or children's, as distinguished from one that is adjusted to the ankle by buttons, buckles, rubber goring, etc., and from the "Blucher," "Polish," "Oxford," etc.

BALL—of the foot; the fleshy part of the bottom, back of the toes.

BEADING—Folding in the edges of upper leather, instead of leaving them raw.

BELLOWS TONGUE—A board tongue sewed to the sides of the top, as in waterproof and some working shoes.

BLACKBALL—A mass of grease and lampblack, formerly used by shoemakers on edges of heels and soles; sometimes called "cobbler's botch".

BLUCHER—A shoe or half boot, originated by Field Marshal Blucher of the Prussian army, in the time of the first Napoleon. It at once became very popular, and has since received occasional favor, being used with high tops as a sporting or hunting boot. Its distinguishing feature is the extension forward of the quarters, to lace across the tongue which may be an extension upward of the vamp.

BOOT—The term is sometimes used (especially abroad) to designate ladies' high cut shoes. Here it applies only to high topping footwear, usually made with the tops stiff and solid, sometimes laced, as in hunting boots.

BOOTEE—A boot with short top or leg, usually made with rubber goring over the ankle, sometimes with a lace front to imitate the appearance of a shoe in wearing.

BOXING—Stiffening material placed in the toe of a shoe to support it and retain the shape; leather, composition, zinc, wire net, drilling stiffened with shellac, etc., are used.

SOFT TIP—A term applied to a shoe on which no underlying boxing is used under the tip.

BROGAN—A heavy pegged or nailed work shoe, medium cut in height.

CACK—An infant's shoe made with a sole leather bottom without a heel.

CHANNEL STITCHED—A method of fastening soles to the upper, either by the McKay or Welt process, in which a portion of the sole on the outer side is channeled into and the stitches afterwards covered on the lower side by the lip of this channel.

CAP.—Same as tip.

CASE—of shoes; so far as quantity is concerned, the contents of cases of different shoes vary. Men's boots and shoes usually come twelve pairs to the case; women's twenty-four to thirty-six pairs; children's as high as seventy-two pairs. Cases for foreign shipment are made much larger. Estimates of shipments are usually based on the number of cases, one reason for this being that it is much easier to tabulate the shipment from any point by collecting reports direct from the freight houses, where the handlers count the packages, with no means of knowing the precise contents.

CHANNEL SCREWED—A process by which the sole is fastened to the uppers in the following manner: After a channel is cut and laid over on the outside of the outsole, the outer sole and inner sole are fastened together holding the upper and lining between them by means of wire screws which are fastened in this channel. The skived part is then smoothed down over the heads of the screws entirely covering them from sight and affording a means by which the screws are not easily worked up into the foot.

COLONIAL—A ladies' low shoe, with vamp extended into a flaring tongue, with a large, ornamental buckle across over instep. The buckle and tongue as distinctive features, whether the shoe fastens with lace or strap.

COMBINATION LAST—One with a different width instep than the ball. It may be one or two widths difference, such as the C ball with an A instep. Combination lasts are as a rule used to fit abnormally low insteps.

CONGRESS GAITER—A shoe with rubber goring in the side which adjusts it to the ankle, instead of laces, etc.

COUNTER—The piece of stiffening material that passe. around the heel of the foot to support the outer leather and prevent the shoe from "running over" at the heel. It is made of sole-leather, shaved thin on the edge and shaped by machinery, in the best shoes. Is made of composition or paper, in cheap shoes, and metal is occasionally used, sometimes on the outside of the shoe in heavy goods for miners and furnacemen.

CREEDMORE—A heavy, men's lace shoe, with gusset, blucher cut.

CREOLE—A heavy, congress work shoe; these and creedmores, brogans and "Dom Pedros" are usually made of · oil grain, kid or split leather, sometimes pegged, sometimes "stitched down".

CUSHION SOLE—An elastic inner sole (proprietary).

DOM PEDRO—A heavy, one-buckle shoe, with gusset or bellows tongue. Originally a proprietary name, and shoes so-called were made of fine material, now usually in cheap grades.

EYELET—A small ring of metal, etc., placed in the holes for lacing; the eyelet-holes are sometimes worked with thread like a button hole.

FAIR STITCH—Stitching that shows around the outer edge of the sole, usually to give a McKay the appearance of a welt shoe.

FORM—A term applied to a filler last. They may be of wood, papier mache, leather board or any material and are used to enhance the appearance of sample shoes, either in salesmen's lines or in retail window displays.

FOXED—Having the lower part of the quarter a separate piece of leather or covered by an extra piece; "slipper foxed," a term sometimes applied to ladies' full vamp shoes.

FOXING—That part of the upper that extends from the sole to the laces in front and to about the height of the counter in the back, being the length of the upper. It may be in one or more pieces and is often cut down to the shank in circular form. If in two pieces, that part covering the counter is called a heel fox.

FRENCH HEEL—(See "heel".)

FRENCH SIZES—(See "size".)

GAITER—Usually applied to a congress shoe, sometimes to separate ankle covering.

GOODYEAR WELT—(see welt".)

HEEL—Varieties; "Cuban heel," a high straight heel, without the curve of the "French" or "Louis XV" heel, which is of extreme height, and is thrust forward under the foot with curved outline in back and "breast," or front surface; it is sometimes made of wood covered with leather, with "top-lift," or thickness of sole leather, sometimes of all sole leather; "military" heel, a straight heel, not so high as a Cuban; "spring" heel, a low heel, formed by extending back the outsole of the shoe to the heel, with a "slip" inserted between outsole and heel-seat; "wedge" heel, one somewhat similar to a spring heel, except with a wedge shaped lift tacked on the outside instead of a slip.

HEEL PAD—In the manufacture of shoes is a small piece of felt, leather or other substance, fastened to and covering the full width of the insole at the point upon which the heel rests. A heel cushion is sometimes called a heel pad.

HEEL SEAT—The rounded hollow of the heel of a shoe, formed by use of the "rand" and the curving under of the counter.

INLAY—A trimming of the upper by an insertion of the same or different kind of material than that of the body of which it is inlaid. It is used for decorative purposes on a shoe.

INSOLE—The inner sole, to which in "McKay" and "welt" shoes the upper and the outsole are sewed or nailed. It is an important though invisible portion of the shoe, and should be of the best leather. It is the foundation of the shoe.

INSTEP—The top of the arch of the foot.

IRON—A term indicating thickness of sole leather, it is one thirty-second of an inch.

JULIETTE—A women's house slipper which is cut a little above the ankle in front and back and is cut down on the sides. They are usually fur trimmed. The same style in kid with goring in sides is called a "Juliette" by some manufacturers.

LACE-STAY—The strip of leather reinforcing the eyelet holes.

LACE HOOK—An eyelet extended into a recurved hook, around which the lace is looped; most commonly used in men's and boys' shoes, although recently some have been invented for use in ladies' shoes, with curved in ends, to avoid catching the dress.

LAST—The wooden form over which the shoe is constructed, giving the shoe its distinctive shape. The lasts are a large item of expense in shoe manufacturing, as each different shape of shoe requires a complete set of lasts for each size and half size made. They are turned from seasoned maple wood, and are often made with a hinge to facilitate pulling them out of shoes. Those used for "McKay" work are metal bottomed.

LIFT—On thickness of the sole leather used in making a heel; "top" lift is the bottom lift, when the shoe is right side up; the last piece put on in manufacture.

LINING—Usually made of some form of drilling or sheepskin. Often a nuisance to the wearer, when improperly fitted. The lining in the best shoe is as carefully cut and fitted as the upper. Dealers will find that customers vary greatly, in the matter of wear of linings. Some will "go through" the linings in short order, and be back in the store with complaints. In this as in other things, the best is usually the cheapest.

McKAY SEWED (or simply "McKay"); a shoe in which the outsole is attached to insole and upper by a method named for the inventor. The upper is lasted over an inner sole, the last removed, and the outsole sewed on by a thread that goes through from the outside, catching upper and inner sole, and leaving the seam showing on the inside. Before sewing, a channel is cut and laid over on the outside of the outsole, and afterward pasted back over the seam as in welts. A "sock-lining" is usually put in the shoe to cover the inside seam. It is a cheaper method than welt sewing, but makes a shoe less flexible and less susceptible of repair. This and the "turned" and "welt" (defined elsewhere) are the three principal varieties of shoes in which the soles are attached by sewing, instead of pegging, nailing, etc. The McKay sewing method accomplishes in one operation what the welt method requires two for, namely, the attachment of upper, insole and outsole. Its introduction cheapened the making of medium priced shoes, although improvements have been made in welt methods recently that are greatly strengthening their standing in the total of shoes made.

MOCK WELT—McKay sewed shoe with a double sole and having a leather sock lining.. It is fair stitched to imitate a welt.

MULES—Slippers with no counter or quarter.

NULLIFIER—A shoe with a high vamp and quarter, dropping low at the sides, made with a short rubber goring, for summer or house wear.

OXFORD—A low cut shoe, no higher than the instep, lace, button, or goring, made in men's, women's and children's sizes.

PACS—Coverings for the feet made of good quality calf-skin similar in form and appearance to the Indian moccasin. They do not have sole leather bottom. If properly made, they are waterproof. Used sometimes in plowing in soft ground.

PASTED COUNTER—A counter cut from two pieces of sole leather pasted together. Sometimes called two-piece counter.

PATTERN—The models by which the pieces, composing the upper of a shoe are cut; applied collectively to the upper as modified by the differing shape of these pieces. .

POLISH—A ladies' or misses' front lace shoe, of higher cut than a "bal," named from Poland, where it is said to have originated, and pronounced accordingly.

PUMP—A low cut shoe originally having no fastening such as laces or buttons. Recent patterns, however, are being made with one eyelet on either side or straps. A pump is cut lower than the instep.

PUMP SOLE—An extra light single sole running clear through to the back of the heel, a Yankee trade term for single sole McKay. A pump sole in former years was distinguished by its flexibility and was hand-turned. The present day term, however, is applied to extra light weight soles; being lighter weight than the regular single sole and are usually seen on men's heavy work shoes.

QUARTER—The back portion of the upper of a shoe covering the counter and extending forward, containing the lace eyelets.

RAND—A piece of leather at the top of a heel, extending around the heel under the sole, with the inner edge made thin, so as to form with the curved counter a rounded inside for the heel of the wearer to rest in.

SABOT—A one-piece wooden shoe, carved from a block of bass-wood. A novelty to most localities, although many thousands are worn by foreigners in this country. France and Germany both claim the honor of originating it.

SANDAL—A ladies' strap slipper; originally applied to a sole fastened on the foot by thongs or straps, of ancient use.

SCREW-FASTENED—Having the sole attached with screws, as in cheap or working shoes.

SHANK—The middle portion of the bottom of a shoe that comes under the arch of the foot; "Shank steel," a curved piece of steel built into the shank to support it.

SLIP—Applied to spring heels or to soles; a thin piece of sole leather, inserted above the outer sole.

SIZES—The smallest shoes, "infants," run from 1 to 5; then "children's" in two series, 5 to 8, and 8½ to 11; then they branch out into "youths'" and "misses'", both running 11½, 12, 12½, 13, 13½ and back again to 1, 1½ and 2, in a new series of sizes that run up into "men's" and "women's"; "boys'" shoes run from 2½ to 5½ "men's" from 6 to 11, in regular runs. Larger sizes are, of course, made, but only upon special orders. Some few manufacturers go to 12, but not many. Women's sizes run from 2½ to 9. Some makers do not go above 8's. The scale of sizes is sometimes varied from by manufacturers of specialties. The "little gents'," usually from 10 to 13½, is an incidental run of sizes. Size No. 1, in infants', is (or was originally) four inches long; each added full size indicates an increase in length of one-third of an inch. A man's No. 8 shoe, therefore, would be about 11 inches long. These measurements are not now absolute. They were originated in England. What is known as "French sizes" refers to a cypher system of marking to indicate these sizes as well as widths, so that the real size need not be known to the customer. The GAZETTE inclines to believe that American ingenuity is too modest in giving this useful idea a foreign label. An example of a French size system, in actual use from the catalogue of a well known firm, is here given:

Widths......	AA	A	B	C	D	E	EE
Mark	9	10	11	12	13	14	15

Sizes........	1	1½	2	2½	3	3½	4	4½	5	5½	6	6½	7	7½
Mark	20	20-	21	21-	22	22-	23	23-	24	24-	25	25-	26	26-

In using this system, instead of having widths marked A, B, etc., the number just below the letter in the above key is used. The same with sizes, so that, for instance, a No. 6 shoe, D width, would be marked simply "1325." The second figure, and the fourth, are the ones for the salesman to remember. The second figure, 3, is one less than the number of the size letter, D, which is the fourth letter; and the fourth figure in the mark, which is 5, is likewise one less than the correct number of the size of the shoe, 6. The mark "1325" would convey no information to the customer.

SOLES—Varieties and modifications; a "full, double" sole, has two thicknesses of leather extending clear back to the heel; "half double" sole a full outer sole with slip extending back to shank; "single" sole, is self-defining; "tap" is a half sole. Materials are defined in a separate article on leathers. Methods of attaching are discussed under "McKay," "turned," "welt," etc.

STITCH-DOWN—One of the simplest of all shoes in construction, in which the top is turned out instead of under, and stitched down through the sole, as the sole on a welt shoe is attached to the welt. It is flexible shoe, that was largely used in the army, and is still worn by many veterans.

STITCHED ALOFT—A term applied to a method of fastending shoes by the Goodyear welt system by which the stitching shows on the bottom side.

STRAIGHT LAST—One that is neither right or left and a shoe made over such a last can be worn on either foot. This term is sometimes applied to right and left shoes that have a barely perceptible outside swing.

SWING—A term. applied to the curve of the outer edge of a sole.

TIP—An extra piece covering the toe, separate from the vamp; "stock tip," a tip of the same material as the vamp; "patent tip," a patent leather tip; "diamond tip," refers to the shape, extending back to a point; "imitation tip," stitching across the vamp in imitation of a tip.

TOP FACING—The strip of leather or band of cloth around the top of the shoe on the inside. It adds to the finish and beauty of the lining and is sometimes used for advertising the name of the manufacturer or retailer by having a design of letters woven or sewed in or on it.

TURNED SHOE—(or "turn" shoe); a ladies' fine shoe, that is made wrong side out, then "turned" right side out, which operation necessitates the use of a thin, flexible sole of good quality. The sole is fastened to the last, the upper is lasted over it wrong side out, then the two are sewed together, the thread catching through a channel or shoulder cut in the edge of the sole. The seam does not come through to the bottom of the sole where is would chafe the foot on the inside. If you have a new cobbler be sure that he understands the construction of a turn shoe before he tries to repair one.

UPPER—A term applied collectively to the upper parts of a shoe.

VAMP—The front part of the upper of a shoe; "cut off" vamp, one that extends only to the "tip", instead of being continued to the toe and lasted under with the tip; "whole" vamp, one that extends to the heel, without a seam. The vamp is the most important piece of the upper, and should be cut from the strongest and clearest part of the skin.

VESTING—A material originally designed, as its name would indicate, for making fancy vests. As used in shoes, it is made with fancy figured weave, having a backing of stiff buckram or rubber treated tissue to strengthen it.

WELT—A narrow strip of leather that is sewed to the upper of a shoe, with an insole, leaving the edge of the welt extending outward, so that the outsole can then be attached by sewing through both welt and outsole, around the outside of the shoe. The attaching of the sole and upper thus involves two sewings; first the insole, welt and upper, then the outsole to the welt. The name is applied to the shoe itself when made in this way, as distinguished from a "turned" or "McKay sewed" shoe. This was (and is)the method used by cobblers, in the production of hand sewed shoes, to fasten together the sole and upper; "Goodyear welt," a welt shoe in which the sewing is done by machines, named for the inventor, the Elias Howe of the shoe world. There are very few hand-welted shoes made, although many are so designated.

WIDTHS IN RUBBERS—A complete list is: S, Slim; N, Narrow; M, Medium; F, Full; FF, Extra Full; W, Wide; WW, Extra Wide. (For width of shoes see sizes.)

Leather Terms Defined.

Terms Commonly Used in the Retail Trade, Revised and Brought Down to Date.

The following list of terms includes many that are commonly heard in the boot and shoe trade, and the definition and description offered for each is believed to be correct. No attempt has been made to include all, or even any considerable part, of the proprietary names given to various kinds of leather. There are few cases familiar to all, in which the trade name has become a generic term by general usage.

ACID TANNED—(See "Tanning.")

BELTING—Usually bark tanned cowhide, used in various thicknesses for machinery belts. Clear stock, of firm texture, is required.

BLOOM—(See "Spewing.")

BOX CALF—A well known proprietary leather having a grain of rectangularly crossed lines.

BUFF—A split side leather, coarser than "glove grain," but otherwise similar. It is used for cheaper grades of shoes, principally men's. See "split."

CARBARETTA—Tanned sheepskin of superior finish used for shoe stock. It was for a time alleged to be the skin of a mythical animal, a cross between a goat and a sheep. There are sheep with wool not far removed from hair in texture, which produce a skin of greater tenacity and finish than the ordinary sheep.

CALFSKIN—Skins of neat cattle of all kinds, from Texas long-horns to the sacred cattle of India, weighing up to 15 pounds, are usually included under this term. It is simply an arbitrary distinction, for trade convenience. They make a strong and pliable leather, that has been used for many years for men's shoes and heavier grades for women's shoes. The next grade in weight, from 15 to 25 pounds, are called "kips". Men's boots are made from these, as well as from heavier hides split in two or more layers. (See "Splits.") Calfskins were formerly always finished with wax and oil on the "flesh" side, but can now be made so as to be finished on the "grain"—the

101

hair side of the skin—the outside, in which show the markings of texture caused by the pores, etc. Bark tannage has in a large measure been superseded (not entirely) by the chrome process.

CHROME TANNED—(See "Tanning.")

COLTSKIN—The firm texture of horsehide has brought it into more general use in shoe making within the past few years. The skin of a colt is thin enough to use like calfskin in its entirety, with such shaving as is given all hides in tanning. With the hides of mature animals, splitting is resorted to. Coltskin makes a firm basis needed for patent leather, and has been much used in recent years for this purpose. Russia is the chief source of supply.

COMPOSITION—A variety of methods are used to work up the small scraps that accumulate about tanneries and factories. Some of it is ground up and mixed with a paste or kind of cement and flattened into sheets, which are used as insoles and in other parts of various grades of shoes, where wear is not excessive.

CORDOVAN—A name long in use, originally signifying Spanish leather, made from either goatskin or horsehide. The Spaniards were the best leather makers for centuries. As used now, the term is applied to a grain split from the best and strongest part of a horsehide.

COWHIDE—Used to refer to hides of cattle heavier than "kips", which run up to 25 pounds each.

DONGOLA—Heavy, plump goatskin, tanned with a semibright finish.

ENAMEL—Leather that is given a shiny finish on the "grain" side, as distinguished from "patent leather," which is usually finished on the flesh side, or the surface of a split. The process is similar. See "patent."

GLAZED KID—(See "Kid.")

GLOVE GRAIN—A light, soft finished split leather, for women's or children's shoes or topping.

GOATSKINS—(See "Kid.")

GRAIN—(See under "Calfskins.")

HARNESS LEATHER—Similar to belting, and made from hides heavier than kips.

HIDES—As distinguished from "skins," under which term are described the skins of goats, calves, sheep and other small animals, hides refers to those skins of cattle which are above 25 pounds in weight; also to skins of horses, etc. It is a trade distinction only.

HEMLOCK TANNED—(See "Tanning.")

KANGAROO—The skin of this antipodean marsupial makes splendid leather, of firm texture. But there are kangaroos by nature, and kangaroos that become so by postmortem brevet in the tannery. The genuine is quite expensive.

PACKER HIDES—Are those taken off in the large slaughtering houses. They are rated slightly higher in price, because of the greater care and skill used in taking them off.

KID—This term is applied to shoe leather made from the skins of mature goats. The skin of the young goat or kid is made into the thin, flexible leather used in making kid gloves, being too delicate for general use in shoes. The goats from which comes the supply of leather used in this country for most ladies' fine shoes, many children's shoes and an increasing number of men's shoes, are not of the common domesticated kind known in this country, but are wild goats or allied species partially domesticated, and are found in the hill regions of India, the mountains of Europe, portions of South America, etc., The process of tanning is naturally quicker than the tanning of heavier hides, and all varieties of tannage are used, the chrome methods having come into very general use. There are many kinds of finish given, such as glazed, dull, matt, patent, etc. One quality that distinguishes goat leather, the "kid" of shoe making, is the fact that the fibers of the skin are interlaced and interlocked in all directions. Instead of ripping straight through, like a piece of cloth, or splitting apart in layers, as sheepskin will do when made into leather, the kid holds together firmly in all directions. The finished skins as they come from the tannery, by whatever process they may be put through, are sorted for size and quality, a number of grades being made. This sorting is repeated in the shoe factory, sometimes repeatedly, the effort being made to secure in each lot or grade of shoes made as nearly perfect in uniformity of texture and quality and weight as possible. This is a work of importance, and one requiring good practical knowledge of leather. No two skins come through exactly alike. They vary as to fineness, thickness and size, some are scarred by careless skinning; some have had spots caused by careless handling before they reach the tannery; some are given too strong a dose of chemicals in the tanning itself; all these things are considered in all the many sortings that take place in their progress through the tannery and the shoe factory, from the raw skin to the finished product, regard being had to the quality and kind of shoe it is desired to produce. (The same sorting, it may be said, is done with all leathers.)

KIP—(See Calfskin.)

MATT—A term applied to dull-finish kid, as distinguished from "glazed," etc.

MOROCCO—Originally a leather made in the country of that name, a sumac-tanned goatskin, red in color, such as is still made there and in Europe and used for book-binding.

The name is applied also to leather made in imitation of this, and in general to heavy, plump goatskin, used for shoes.

MONKEY SKIN—This has a peculiar grain, and is among leathers that may be classed as fancy. It is often imitated.

OAK-TANNED—(See Tanning.)

OOZE—A chrome tan calfskin treated on the flesh side in such a manner that the long fibers are loosened and form a nap surface. Made in many colors.

PANCAKE—One of many "artificial leathers," so-called, made from leather scraps, shaved thin and cemented together under heavy pressure.

PATENT LEATHER(Kid, calf, etc.)—There is no longer any patent on the principal processes that are required to make what is known as patent leather, which might be described with fair accuracy by calling it japanned or varnished leather. Calfskin is shaved on the flesh side to uniform thickness, and successive coats of liquid black varnish are applied, the first coats being dried and rubbed down, so as to work the liquid thoroughly into the fiber of the leather. The last coat is applied with a brush, and is allowed to dry in direct sunlight, which seems to be essential. Various formulas are used in making the varnish, vegetable gums and oils forming important ingredients. Like any other such coating, it is liable to crack. The really conscientious dealer can always guarantee patent leather shoes to do that sooner or later, if worn. Kid and coltskin have been largely used during the past year as a basis for patent leather. The former is more elastic, and it seems possible to give it some degree of porosity, thus removing one serious objection to the use of patent leather for shoes, namely: its air-tightness—an objection urged by consideration of both hygiene and comfort.

SATIN CALF—A grain split, stuffed with oil and smooth finished.

SEAL GRAIN—Usually a flesh split, with an artificial "grain" or kind of indented tracery, which is stamped or printed on the finished leather.

SHEEPSKIN—Used largely for linings and for cheap shoes for women and children. It is too soft and weak in texture for heavy wear, and is liable to split and tear.

SKINS—(As distinguished from "hides."—See Hides.)

SOLE LEATHER—Made from the heavier hides of cattle, and tanned with oak or hemlock bark (various other vegetable extracts sometimes being used). The oak-tanned is preferred, and may be known by its light color. The hemlock-tanned is of a red shade. In "union tanned" hides, both oak and hemlock are used, and the result is a compromise in both color and quality.

SPEWING—Shoes in stock sometimes become coated with a grayish white powdery substance, that looks like mildew. This formation on leather that is not fully seasoned is called "spewing", and the deposit is called "bloom." It can readily be wiped off, and does not indicate any serious defect or trouble with the leather. It is not a mildew or growth, but apparently an exudation of materials used in tanning.

SPLITS—Are what the name implies, split leather. A thick hide, often before the tanning process is completed, and after being cut in two down the back into "sides," is run through a machine between rollers and impinging upon a sharp knife edge, that splits it into two or more sheets. The knife used might be described as a flexible razor three feet long, and must be kept perfectly straight and as keen as it can be made. Its care and use requires a considerable amount of skill, and tanners say that to teach a new splitter the trade costs them at least a thousand dollars in damaged and imperfectly split leather. A split from a heavy hide is, of course, not as good as the whole of a lighter hide, light enough to make leather of the thickness required, without splitting. Hides above the weight of "kips" (25 pounds) are used, ordinarily, and the leather that is intended for use in shoes receives various finishes and is known by various names, such as seal grain, buff, glove grain, oil grain, satin calf, russet, plow shoe, etc.

SUEDE—A trade term applied to kid skins, finished on the flesh side. Some leather makers have applied this term to velvet finished calfskins on the grain side, which is not regarded as entirely correct. Made in many colors.

TANNING—It is impossible to give in small space a description of all the details of tanning, but the basic principle is the treatment of the raw skin with tannic acid or some similar substance having an astringent of "puckery" effect. The hides or skins, of whatever kind, are first thoroughly soaked and cleaned, large ones usually being cut in two up the back into "sides" for convenient handling. The next step is the removal of the hair, which is effected either by a sweating process, the hides being hung up in warm rooms and kept there almost to the beginning point of decomposition, or by soaking in a solution containing lime. As soon as the hair follicles are sufficiently dissolved by either process as to permit of the hair or wool being easily pulled or scraped off, the hides are worked through machines or by hand. Further soaking in "vats" of solutions to counteract the lime follows, and the skins are ready for the tanning proper. In the methods that involve the use of oak or hemlock bark or other vegetable extracts, the hides are placed in vats of solution of varying strength, and worked back and forth for a number of weeks. Upon completion of this stage, the stock receives its final dressing and is stuffed with oil or grease, dyed black, polished, etc., according to the results desired. Sole leather is oiled but slightly and is dried and rolled smooth. The "chrome" process involves the use of a salt of chromium usually bichromate of potash, with muriatic acid. This process is very quick in action, taking but a few hours for the tanning, the hides being prepared as for the bark tanning. Both goatskins and calfskins are tanned by the chrome process, making a strong, durable leather, with considerable water resisting power. Every detail of all processes involves care and experienced skill, and there is an infinite variety of finished products from the highest glove or fancy leather to the heaviest sole leather. Chrome tanned leathers are given a comparatively dry finish, although some oil, dissolved in acid, is used.

TAWING—Making leather by soaking hides in a solution of salt and alum, or by packing down with dry salt and powdered alum—essentially a tanning process. It is used to prepare skin rugs and furs.

VELOURS—French for "velvet." A trade name for a proprietary chrome tanned calf. The leather is of smooth and velvety finish and excellent quality.

VICI—A proprietary trade name for a brand of chrome tanned kid, which has almost become a generic name, by common use and sometimes not intentional misappropriation.

VISCOLIZING—A proprietary method of water-profing sole-leather, evidently by the use of some partly emulsified oils, with a water resisting tendency. The viscolized soles are used in hunting and sporting boots, the method softening the leather to some extent.

Shoe Stock Keeping.

Simple Forms of Records Are Always the Best.

How should one keep the shoe stock?

The answer is simple. Find the least difficult method of accurately recording three things—goods coming, stock on hand and shoes sold.

Use only the details necessary for your own particular business. The real purpose in stock keeping is to keep the stock at the lowest possible level, to eliminate the sizes and styles not needed and at the same time to keep on hand those sizes and styles in greatest demand. In other words stock keeping of the right sort means greater profits.

All shoe stores and shoe departments use some stock keeping plan. Some stores go too deeply into details and others do not give nearly the attention they should have along these lines. The purpose of this article is not to describe any one plan but to give a basis which can be applied in various plans to any stock.

Three Divisions of Stock.

In the first place there are three divisions of the stock which must be taken into consideration in any shoe stock. First the goods now on the shelves; second the sizes and styles ordered but not yet received; third the shoes which have been and are being sold.

The buyer must know what sizes and styles have been purchased or he is likely to duplicate the order with similar styles or sizes. This is the cause of a great deal of the stock accumulations. It is necesasry to know what is on hand and to know all about these goods or there will be duplication in buying and forgetfulness in pushing the slow sellers at the height of the selling season or in ordering the sizes which are needed right now. Everything must be known about the goods sold to anticipate the requirements of the people in the community. or it will not be possible to judge what to buy in the future or to anticipate the requirements of the people in the community.

Two Methods of Keeping Records.

There are two means of recording these things or any part of them. One is a stock book with ruled pages, one for each style of shoe or a card index with a card for each style of shoe. In the latter case no dead wood is carried in the record. When a style is completely sold out the card record of that style can be destroyed without affecting the other records. In a stock book the page must be carried and if the book becomes filled then a complete copy of all the records often has to be made. A stock card either of one's own design may be used or one which can be purchased ready ruled for this purpose. The card should have room on it for all the sizes and widths and enough spaces for all the pairs which will ever be sold of any shoe. These cards can be bought with as many as fifteen hundred spaces on them.

The three divisions of the stock can be indicated by arbitrary marks. A straight line for goods coming, a cross for goods on hand and a third line drawn through the other two or a circle drawn around them to indicate the goods sold. Colored ink may be used if desired. Black ink lines for the first division, red ink lines for the second, and penciled lines for the third. The latter can be erased from the records if the shoes are returned for any reason.

Means of Gathering Records.

Nearly all systems are based on the above plan. The greatest difference in stock keeping is the means of gathering the records. Some stores save the duplicate order blanks to get the records of the first division. Others use the simpler method of entering the sizes on stock cards and filing them in the same cabinet with all other styles records. . Some stores use sales checks and require the salespeople to write the stock number of the shoes, the size, width and often the name of the customer on the check. This, however, takes time, and frequently when it is needed for actual selling. Another plan is to have small tickets on each carton on which is printed the size, price and stock number. Then all that is necessary is to turn in the ticket and fill in the card records at leisure.

The next important feature of any stock keeping system is the stock numbering plan. It is necessary to have a stock number for each different style of shoe. It is better to use a plan by which the stock number will describe the shoe. It is not necessary, except for harmonious appearance, to use special labels. The stock number can be imprinted on the manufacturer's label with an ordinary band stamp.

Stock Numbering Plan.

Any sort of stock numbering plan which will accomplish the desired results can be used. We suggest to those who have no special one the following:

Mark all men's shoes between 1 and 999; women's shoes between 1,000 and 1,999; boys' shoes between 2,000 and 2,999; girls' shoes between 3,000 and 3,999; rubbers and findings between 4,000 and 4,999.

Subdivisions for each of the above: High shoes between 1 and 500; oxfords between 500 and 999; patent leathers between 1 and 99; kidskin leathers between 100 and 199; dull calf leathers between 200 and 299; bright leathers between 300 to 399; other leathers between 400 and 499.

Odd ending numbers for welts and double soles and even ending numbers for turns and light McKays.

For colored shoes add fractions; ½ indicates tans; ¼ indicates red; ⅛ indicates blue and so on.

Examples: 333 indicates men's box calf boot with double sole; 833 indicates men's box calf oxford with double sole; 3,011 indicates girls' patent oxfords, turn sole; 1,255 indicates women's gun metal welt boot; 1,254½ indicates women's tan calf turn boot; 1,942⅛ indicates women's blue silk turn pump.

The Shelf Arrangement.

The next feature of importance is the shelf arrangement. Shoes should be arranged in such a manner that no time is lost in locating any pair. There are three good plans for three different kinds of stock.

In small stores or in general stores where the clerks sell all two widths, a good plan is to arrange the shoes on the shelves over the store and where the shoes are carried in but one or according to sizes. All the two and one-half in women's shoes can be carried together with the cheapest grades first, followed by the next better grade. The shoes can be run from left to right clear across a section. The small sizes will be near the ledge and the large sizes at the top of the shelving. The clerk can thus always find all shoes which will fit any one person in a certain spot in the shelving. It makes selling easy to those not very familiar with the stock.

In stores of medium size, carrying the average stock, the lot plan of stocking the shoes on the shelving is a good one. This plan consists of placing all the pairs of any certain style together on one shelf. Any plan of grouping will answer.

The larger stores with their shoes carried in widths, find the up and down plan to be the best for their business. Start the smallest size and the narrowest width first with the next larger size or width above it. The continual shifting of the stock gives each · style an equal chance with the others. One lot may be at the top of the shelving one week and in easy reach the next.

The best stock system is the simplest one. Any plan which will save work in the stock, which will keep track of the stock or place it where the salespeople can find it more easily is the thing to use. While this article covers the ground only in a general way we would be glad to go into details upon any subject of shoe stock keeping. If you feel that your stock is not arranged just right, if you want to keep a better record of the three divisions or if you wish to improve or change any feature of your present plan write The Shoe & Leather Gazette the details of your system and it is more than likely that we can suggest the best plan for you.

The Reading Population of the U. S. A.

knows things to-day about shoemaking which have for generations been a part of the " Mysterious Art" of making shoes. They know how a shoe can be made that is perfectly smooth inside, therefore comfortable; made in a manner that requires good materials, therefore durable. In fact,

The Whole Story of the Goodyear Welt Shoe

has been told in the leading publications of the country, and the public interest has been manifested in the flood of requests which we have received from every section of the world for the interesting booklets we are pleased to send to all who desire them. The demand for Goodyear Welt Shoes is a perfectly logical one; to know them is to want them. You can't say anything to your customers that will so strongly impress them with quality as to say " Goodyear Welt ", and it is well to say it to them before they say it to you.

United Shoe Machinery Co.
Boston   Mass.

How to Make Profit.

A Fair Margin Is One-Half More Than the Cost of the Shoes.

In this day of advancing prices on practically everything which is bartered and exchanged just what profit belongs to the retailer when he sells a pair of shoes? This question put to 100 retailers will probably bring forth a variety of answers from—"Sell your shoes as close as possible", to "Get all you can". Still, there is an average profit which all merchants should get and which we believe a whole lot of merchants are not getting.

Obviously all merchants are in business to make all they can, but competition compels them to sell their wares at reasonable prices. It sometimes happens that all merchants in the town mark their goods at a fair margin of profit above expenses. In such cases all make money. On the other hand, there are some towns in which one merchant offers shoes at a price by which he makes a gross profit, but actually loses money when his real expenses are figured in. If all the other merchants meet his prices, then they are bound to lose money.

There is a limit at which good merchants should stop. Just what that limit is can be estimated from the results of the following investigations. We found that in three different size towns of population in the following order, 750,000, 50,000 and 20,000 that the percentage of doing business was respectively: 27½, 20 and 18 per cent. It has been stated by commercial authorities that the average shoe store is at an expense of 20 per cent on the selling price.

We are showing a table of figures indicating at what price shoes should be marked if 25 per cent on the selling price is obtained and if 33 per cent on the selling price is obtained. Do not confuse 25 per cent on the selling price with 25 per cent on the cost price, because 25 per cent on the selling price will figure 33⅓ per cent on the cost price. For example: a shoe which costs $2.25 a pair and sells at $3.00 is marked at 25 per cent on the selling price and 33⅓ per cent on the cost price.

Occasionally we find the retailer who is selling shoes which cost eighty cents for a dollar a pair. If it costs him twenty per cent to do business then he must sell every pair of every lot, and have no complaints or adjustments in order to come out even. The chances are that one or two pair will be sold

111

Wholesale Cost	MarkStaples 33% on Cost or 25% on Selling	Mark Novelties 50% on Cost or 33⅓% on Selling
....$0.37½	....$0.50	....$0.56¼....
.... .50	.... .66⅔....	.... .75
.... .60	.... .80	.... .90
.... .75	.... 1.00	.... 1.12½....
.... .85	.... 1.13⅓....	.... 1.27½....
.... 1.00	.... 1.33⅓....	.... 1.50
.... 1.10	.... 1.46⅔....	.... 1.65
.... 1.15	.... 1.53⅔....	.... 1.72½....
.... 1.25	.... 1.66⅔....	.... 1.87½....
.... 1.30	.... 1.73⅓....	.... 1.95
.... 1.35	.... 1.80	.... 2.02½....
.... 1.40	.... 1.86⅔....	.... 2.10
.... 1.50	.... 2.00	.... 2.25
.... 1.60	.... 2.13⅓....	.... 2.40
.... 1.75	.... 2.83⅓....	.... 2.62½....
.... 1.85	.... 2.46⅔....	.... 2.77½....
.... 1.90	.... 2.53⅓....	.... 2.85
.... 2.00	.... 2.66⅔....	.... 3.00
.... 2.10	.... 2.80	.... 3.15
.... 2.15	.... 2.86⅔....	.... 3.22½....
.... 2.25	.... 3.00	.... 3.37½....
.... 2.35	.... 3.13⅔....	.... 3.42½....
.... 2.50	.... 3.33⅓....	....: 3.75
.... 2.60	.... 3.46⅔....	.... 3.90
.... 2.65	.... 3.53⅔....	.... 3.97½....
.... 2.75	.... 3.66⅔....	.... 4.12½....
.... 2.85	.... 3.80	.... 4.27½....
.... 3.00	.... 4.00	.... 4.50
.... 3.25	.... 4.33⅔....	.... 4.87½....
.... 3.50	.... 4.66⅔....	.... 5.25
.... 3.75	.... 5.00	.... 5.62½....
.... 4.00	.... 5.33⅓....	.... 6.00
.... 4.25	.... 5.66⅔....	.... 6.37½....
.... 4.50	.... 6.00	.... 6.75
.... 4.75	.... 6.33⅓....	.... 7.12½....
.... 5.00	.... 6.66⅔....	.... 7.50

for less than a dollar and he will be losing money. Now if that merchant knew exactly how much it cost to sell the shoes he would mark them a dollar and a quarter. The shoe merchant who expects to meet present day competition cannot estimate anything; he must actually know all the time.

What is the cost of selling a shoe marked $3.00? What is the most you could pay for it and make a fair profit? How many times do you turn your stock each year and what is the net profit above expenses? These are the sort of questions you ought to be able to answer off hand and know absolutely that you are right.

The three columns of figures in the accompanying table show the basis of two sets of markings which are applicable to the average store.

Column No. 1 can be used on staple shoes which are continually sized in and which are of such a nature that styles do not change. Such shoes as one-strap house slippers; boys' box calf bluchers; old ladies' bals, work shoes and some others, can be marked on this ratio of profit satisfactorily. It seems to us, however, that all novelty shoes which are liable to run out unevenly, or which are suitable for one season only, should be marked according to the second column of figures.

For example: Say that a merchant buys oxfords for $2.00 per pair and that the average number of pairs bought were eighteen. If this oxford seems to be one which will be good for but one season, the chances are that not more than fourteen pairs will be sold before August 1. Four pairs will probably be offered at a discount of twenty per cent, or at $3.40. It may be that one pair out of the lot will prove unsatisfactory and the merchant will be obliged to sell the customer another pair at a concession of one or two dollars, or replace the pair with a new one. Besides, there is the free repair work which is often done in many shoe stores, such as sewing rips, patching, or tacking soles. If the store is under an expense of 20 per cent in doing business, and the merchant figures a 5 per cent profit, in marking them at $2.66 ⅔, the entire lot of 18 pairs will show a loss equal to the cost of the repair work and $1.60 cut price made at the end of the season. On the other hand: If this entire line of shoes is marked $3.00, the net profit obtained on the fourteen pairs will make up for any loss occurring during the season on this particular line of shoes.

$1.60 Shoe at $2.00.

This table should prove interesting to merchants who have paid the manufacturer's advance of 10c a pair on $1.50 shoes and are continuing to sell them for $2.00. At 5 per cent net profit the average store should get $2.13⅓ for this shoe without making any allowance for depreciation of stock and, in order to show a fair profit, the shoe should be marked $2.40. In other words, the shoe merchant who has been selling $1.60 shoes for $2.00 is, in reality, taking 40c per pair out of his pocket. We often hear of merchants who pay $2.35 for a shoe to sell at $3.00. At the very lowest percentage of profit they should obtain $3.13⅔, without allowing for any depreciation and on any shoes which are not strictly staple goods a price of $3.42½, on the average, should be asked by the merchant.

$2.35 Shoes at $3.00.

Merchants who have been selling $2.35 shoes for $3.00 are in reality losing 42½ cents per pair on the price which actually belongs to them. A $2.65 shoe ought not to be marked $3.50, but $4.00. There are, of course, certain staple shoes which can be sold at 25 per cent on the selling price and which will show a net profit of 5 per cent. For example: infants' soft soles at 37½c can be retailed at 50c, as they usually sell out very clean and can be used as a leader. Other lines of shoes marked at 33⅓c will carry enough extra profit to overcome any losses made on soft sole shoes sold at 50c. Some good merchants marked the plain blacks and plain patterns, costing 37½c, at 50c per pair. They mark four or five of the prettier lines, costing the same price, at 60c per pair; thus allowing for any depreciation in faded shoes or small sizes, such as No. 0's or 4's.

$1.00 Shoes at $1.25.

There are still a number of merchants who buy staple house shoes at $1.00 and charge $1.25 for them because they have been in the habit of doing this for years. As a matter of fact, the average merchant loses 8⅓c a pair on every pair of slippers costing $1.00 marked at $1.25. Many times the slipper stock can be averaged up by marking a few of the slippers costing $1.00 at $1.50.

A glance at the table will show that any shoe which is not of an absolutely staple character and which is to be sold at $2.50 should cost in the neighborhood of $1.65. This table of figures is figured on an average business and for an average store. There are some stores in cities, and also in other places which have succeeded in cutting down their expense of doing business to 10 or 12 per cent. The $2.50 shoe stores might be taken as an example of stores doing business on a very small margin of expense. Some of these stores pay as high as $2.10 for shoes to sell at $2.50. They average their business, however, by buying some shoes at $1.50 and $1.65, and by cutting expenses at every turn, manage to make a profit and show good dividends.

In the matter of advertised lines of shoes, on which are stamped the selling price, there should be taken into consideration the value of advertising, which has already been done by the manufacturer. Some lines of shoes which are highly advertised are well known to the public and are much easier sold by the merchant. In other words, he can turn his stock oftener during the year and can make just as much profit by selling on a closer margin than that of his regular lines of shoes. This, we think, is fairly well understood by merchants, generally.

Retail Shoe Advertising.

Suggestions for the Publicity Department of the Business.

There are two ways of looking at advertising, both of them right. Advertising should be done during the dull time for the purpose of starting up trade at that time, and also for the sake of the benefit which will come from being continuously before the public's eye. A man should just as quickly think of stopping his advertising in the dull time, after the holidays, or in the summer as he should think of closing up his store for several months in the year, and keeping it open only when trade would keep him busy. There are probably many houses in the country that could shut up for four months in the year and be ahead in cash at the end of that four months. But, at the end of the next four, the gain would not be apparent. It does not take people very long to forget things, and if the store were closed four months, or the advertising stopped four months, a great many people would have forgotten that the store was in existence.

The other view of advertising is that it ought to be pushed during the busy time when people are ready to buy. Advertising cannot be expected to sell goods when people do not want them, and it will naturally be most effective when it gives publicity to some desirable article at just the right time.

We should think that if a shoe merchant carried a space of four inches double column all the year around, he ought to double the space for the busy months, and occasionally, during that time, he can make larger spaces very profitable.

In business, as in all other affairs of life, everything comes at once. When a business man is so busy with trade and with buying and receiving his goods that he has no time to eat and sleep, just at that time his advertising demands the most careful attention. Just at that time his advertising is the most important part of his business, and usually it is the most neglected part.

In every store somebody has charge of the window display. That seems to be an established rule. The window display is an important part of advertising a store. There are places where, we think, with a good window carefully dressed, such advertising will be all that the store will need. These places are few and far between, however.

Now, if the work of preparing copy for newspaper advertising were turned over to some employe, and a little premium offered him for good work in this line, the result would be a great deal better advertising than is generally done. Of course, all copy should be examined and O. K.'d by the proprietor before it appears in the paper. He will also, of course, decide on what the advertisement was to be about, but this would not take a minute of his time.

If there is no clerk in the store who seems to have an aptitude for this sort of thing, there are a great many professional advertisement writers who could be made use of to good advantage. Some of them are good, some indifferent, and some bad, but most of them will be able to write better advertisements than the merchant can himself, and they will be worth more than they cost in every instance.

The advertisements ought to be changed in every issue of the paper. There are lots of new goods coming in, and each line furnishes material for an excellent advertisement. Publish the ads as if they were news, and tell about the new goods as they come out. If anything comes which seems to be especially desirable, give it particular prominence, and in almost every case it would be a good idea to mention the price.

Whatever you do, do not deal in glittering generalities. Do not put an ad into the paper and say: "John Smith, dealer in footwear. Repairing a specialty. Come here and save money." There is nothing to be gained by such advertising. It probably isn't worth what it cost, but the same space used in an intelligent, thoughtful manner will bring returns every time.

All stores in a given line keep pretty much the same things. The difference between them is made by quality or quantity, or both. People know the generalities without being told. The thing that will attract them is something special and new.

It is just as important for the small shoe dealer to advertise as for the one with the big store. His returns will be just as satisfactory if he does it systematically.

Shoe Window Trimming.

General Tendencies of Shoe Window Trimming Improvement Discussed.

It is an acknowledged fact that a retail shoe business may succeed or fail according to the attention paid the shoe windows. In short, it is the store-keeper's closest connecting link with the public and the class of the store is sized up according to the general appearance of the shop windows. The merchant who considers the goods that he buys must also consider their successful selling. And to sell goods, it is necessary to present them in the most pleasing way—this is the mission of your windows.

The strength of the show window display to tempt the onlooker to enter the store depends largely upon the quality of the display as to its appropriateness at the season. By this we mean that at different seasons of the year different types of display are most effective. At the opening of the spring and fall seasons and at any other event that the store may celebrate by special decorations, elaboration and beauty are the main features.

For high-class showing, the upper portion of the window is rarely used for merchandise display. In most cases, this part of the trim is designed to embellish the display with the use of scrolls, artificial foliage or draperies. As little attention is given to artistic effects on sale windows, it is well to utilize this space for displaying merchandise, rather than let it go to waste. This stocky appearance will also have the tendency to suggest sale merchandise.

The tendency of the leading retailers, especially the department stores, is to place a limited number of well selected shoes in the window at a time. This statement particularly holds good in the showing of new styles at the beginning of the season, although not so many shoes are shown, but what are shown attract more people and they are strongly attracted by the fact that you are showing some mighty snappy footwear styles. By following this plan you can effectively place your merchandise banquet in courses rather than trying to put the whole meal onto the table at once.

The displaying of fewer shoes in the window means that the window should be changed quite often, and this, too, is an advantage in that it continues to attract attention, leaving the impression that there is always something new in the footwear lines to be found at your store.

117

As shoes have such a sameness in appearance and are, as a rule, unattractive in color, it requires a great deal of attention to the background arrangement and construction so as to leave an impression with the public that a change has been made in the display. This naturally brings up the question of suitable background construction. In most cases, in order to secure an efficient show window front, it is necessary to have them boxed from the interior of the store in order to prevent dust and flies from getting into the window and to prevent them from frosting and sweating. A good standard depth for the average show window is 5 feet, which permits of a general showing of footwear. The average height of the window background is about 6 or 7 feet.

The two most popular background finishes of the day are mirrors and hardwood. Both of these, however, have their advantages and disadvantages. Mirrors have one great advantage in their favor, that is the reflecting of all light that comes into the window, thereby making it bright and cheerful. However, it is claimed by some that mirrors have the tendency to confuse the onlooker, especially if the showing is crowded. The cost of mirror background is not much more than that of a good hardwood backing. Its cost depends entirely upon the size of each individual glass utilized. Of course, the larger the glass the better the appearance of the window and the greater the cost. However, some very pretty and effective mirror backgrounds are placed in narrow panels set in hardwood. The hardwood idea background construction is very good for shoes, providing the background is not too dark as to render a display useless in the daytime. This is something that must be guarded against.

Among the varieties of wood used for hardwood window constructions, the most popular are mahogany, oak stained in dark or gray mission effects, and popular stained cherry. When a less expensive background is wanted, high class effects are possible by using some well selected lower priced wood, such as yellow pine or fir. This can be obtained in beautiful grain, and when properly stained and rubbed down to produce a high polish, a very good background can be secured.

In constructing backgrounds of hardwood, one of the panels, usually the center one, is removable and entrance to the window is effected by this means.

Scenic backgrounds are used in window work for special occasions only and are placed in front of the background proper.

One of the most popular background treatments, which will in all probability remain in vogue for some time to come, is the hanging of draperies in appropriate colors from a curtain

118

rod or pleating it at the top edge of the background, allowing the ends to hang in loose, graceful folds to the floor.

This style of background has the advantage of being readily placed in any shape background and the effect can be quickly changed by changing the color of the drapery cloth. It is well to have a number of colors always on hand so as to effect a color change in harmony with the predominating color of the merchandise to be displayed. The best cloth for the purpose is the heavy velour draperies. However, as cheap a material as cheesecloth can be used, if given a backing of white.

Alabastine and burlap backgrounds will appeal to the merchant and the trimmer who must economize in his background expenditure. Alabastine is a dry powder which is mixed with cold water to form a paint. It comes in all colors and shades and has the advantage of being readily mixed in quantities to suit. It is quickly applied with a kalsomine brush. Very satisfactory background effects can be secured by laying out the background in panels of contrasting colors. Burlap in suitable colors makes a good material for the panels. These panels must be set in frames of one inch half round moulding, painted in gold, silver, etc., to give a finished effect.

In using this scheme, the background should be lined with a good weight of sheeting or heavy paper. Asbestos paper is excellent for this purpose and also acts as a fire protection.

Show Card Writing.

How to Begin the Study of Show Card Writing.

The show card is now considered an essential thing in successful retailing. It is a silent salesman which, if properly worded and executed, never offends the prospective customer. The following points are given so as to make this salesman as proficient as possible. Starting with the table, we will follow the show card's list of defects to the window.

Table and Cardboard.

The first thing to consider is the proper location of the work table. If possible, secure space near a well lighted window. A good supply of natural light is much better than electric or gas light, both for the eyes and for good judgment of colors. Next see that the working surface of the table is placed on a slant, as this scheme gives the eye a more direct line on the work, thereby enabling one to space and lay out more correctly.

Don't use cheap, thin cardboard, which is apt to curl in such a manner as to make the wording illegible. Eight or ten ply board is the best for genuine card work. If lighter board is used, it should be mounted or framed so as to hold it in shape. Regular cardboard sheets are 22x28 inches in size. When buying, it is a good idea to have a number of sheets cut with a paper cutting machine, which will produce a clean, straight edge.

Inks and Brushes.

Most of the advertised card writing inks are excellent and you will find them more convenient than mixing your own paint. However, a good paint can be made by mixing dry color with wood alcohol into a thick paste with enough mucilage to make it as thin as syrup, adding a little water at a time.

The best brushes for general card use are red sable riggers. A set of sizes 6, 9 and 12 will answer for all general purposes.

Spacing and Wording.

In spacing, guard against overcrowding and slanting lines, which make a card hard to read. See that the words are properly spaced, well balanced and that ample margin is left to insure an uncrowded and artistic appearance.

The wording is also of most importance. Always aim to let your reader grasp your meaning as he runs. Beware of jokes and slang phrases, as they rarely sell goods; the best show cards give definite information in a forcible and sincere way. Tell your story as briefly as possible; where one word will answer, don't use a dozen. Good wording and catch phrases can be gleaned from various sources, and it is advisable to jot these down for future reference.

Scrolls and Pictures.

Beware of fancy scrolls, loud letters and too heavy shading in gaudy colors, as they have a tendency to give a card a cheap circus poster effect. Study up-to-date, clear lettering—lettering that will answer for practical business purposes, devoid of illegibility in the way of flourishes, shading and gaudy colors. Avoid using pictures of a crude comical character, or drawings and designs that will detract from the wording or appearance of a card. An appropriate picture will often break the monotony which frequent use of black and white cards give. There should be changes in your card schemes, as the public soon tires of sameness.

The first and most important duty of the card is to catch the eye. Neat show card embellishments in scrolls and drawings will help. Next, convey to the mind your argument in forcible words that are strong enough to shift the eye from the card to the display.

Placing Cards in the Window.

Defects are caused by improperly placing the card in the window; show card stands should be used for this purpose. The best stands are those which protect the cards at the corners. The card should occupy a conspicuous place near the center of the display, in plain view of the passing public. The size of the card should be governed by the class of merchandise on display. For example, a small card would be appropriate for shoes, while a large card would be best for furniture. A quarter sheet size, 11x14 inches, is plenty large enough for the average size.

Price Tickets and Soiled Cards.

Price tickets are rarely used on opening displays of high class merchandise. At any time avoid using large price tickets that will hide any of the merits of the merchandise on display, or figures of coarse design in dainty displays of merchandise. Use price tickets in conjunction with show cards in all sale windows, providing the articles on display are of different prices; but if the same price, one show card conspicuously displayed will answer for all.

Mistakes are made by using dirty, torn, faded, ill shaped and long used cards. Nothing is more detrimental to a display of dainty materials than a soiled card. Endeavor always to offer the merchandise in the most pleasing manner consistent with quality and change the cards as often as possible to secure the desired results. By taking pains with old cards, the backs of them can be used again, thereby economizing on your show card expense.

Sizes of Children's Shoes.

How to Find the Proper Size Shoe Without Fitting the Foot.

It is claimed that there are more exchanges of children's shoes than that of adults. In some stores the exchanges on Mondays stand about sixty per cent of children's and forty per cent of men's and women's. The reason for it is that there are so many mothers who buy shoes for children left at home. The exchanging of shoes means a loss to the retailer because it takes time and brings in no profit. Anything he can do to eliminate a part of these exchanges means that much saved.

There are two methods of accomplishing this. One is to instruct the clerks to always inform customers of the advantage of bringing the little ones to the store to be fitted. This can be done in a diplomatic manner so that the sale will not be lost and it will influence the mother to bring the child to the store the next time. The same idea can be incorporated in the advertising or in show cards in the windows or signs in the children's shoe department.

The other method is to have in some convenient place in the children's department a table of average sizes worn at different ages. In connection with this all clerks should thoroughly understand the standard size scale so that the strings, sticks and various measurements brought in by the mothers can be accurately estimated so that the shoes will stay sold and not appear in a few days for exchange.

We are reproducing here a table of sizes which show what the average child will wear at any certain age. These sizes are based on the length measurements and usually apply to D and E widths. If you will notice the table carefully you will see how the rough and ready estimate of adding four sizes to the age of the child will work out.

For example if the mother says the child is five years old add four and you have the average size of nine. If the child is nine years old add four and you have thirteen, the size of the shoe. Or if the child is twelve years old add four and you have sixteen which with the thirteen subtracted gives you size three in little women's. This makes it comparatively easy when the age is known. One point to remember is that you should ask if the child is of normal size or large or small for its age. In the latter cases an allowance should be made, and the size of the shoe figured according to the size of the child and not its age.

It is more difficult when the mothers bring in twigs or strings cut the same length as the child's foot or the outside or inside of its old shoe. But even these can be estimated with a little care so that the exchanges will be cut to the minimum. The story is told of a farmer's wife who came in to buy a pair of shoes for her husband and who when asked the

TABLE OF AVERAGE SIZES WORN AT DIFFERENT AGES.

6 months.............................Average Siz.s	1	— 1½
9 months.............................Average Sizes	2	— 2½
1 year................................Average Sizes	3	— 3½
1 year 6 months....................Average Sizes	4	— 4½
2 years..............................Average Sizes	4½	— 5
2 years 6 months..................Average Sizes	5½	— 6
3 years..............................Average Sizes	6½	— 7
3 years 6 months..................Average Sizes	7	— 7½
4 years..............................Average Sizes	7½	— 8
4 years 6 months..................Average Sizes	8	— 8½
5 years..............................Average Sizes	8½	— 9
5 years 6 months..................Average Sizes	9	— 9½
6 years..............................Average Sizes	9½	—10
6 years 6 months..................Average Sizes	10	—10½
7 years..............................Average Sizes	10½	—11
7 years 6 months..................Average Sizes	11	—11½
8 years..............................Average Sizes	11½	—12
8 years 6 months..................Average Sizes	12	—12½
9 years..............................Average Sizes	12½	—13
9 years 6 months..................Average Sizes	13	—13½
10 years............................Average Sizes	13½	— 1
10 years 6 months................Average Sizes	1	— 1½
11 years............................Average Sizes	1½	— 2
11 years 6 months................Average Sizes	2	— 2½
12 years............................Average Sizes	2½	— 3
12 years 6 months................Average Sizes	3	— 3½
13 years............................Average Sizes	3½	— 4
13 years 6 months................Average Sizes	4	— 4½
14 years............................Average Sizes	4½	— 5
14 years 6 months................Average Sizes	5	— 5½

size said she didn't know but that he wore a sixteen collar. The clerk using common horse sense figured that a sixteen shirt would fit the average man so he sold her an eight E shoe which he later found out was the correct size.

When the mother brings in bits of string which are the length of the child's foot the size of the shoe can easily be found with the aid of a size stick if there is one in the store.

Find the size the twig or string draws on the size stick and add two and one-half sizes and you have the length of the shoe needed. · If you have no size stick handy, measure the string in inches, subtract three and multiply the remainder by three which will give you the size of the shoe needed. For example, suppose the customer brought you a cherry twig seven inches long which she said was the size of the child's foot. Subtract three, which leaves four and multiply by three, which will give you twelve or the size of the shoe needed. To prove it ask her if the child is not about eight years old. The same twig on a size stick would draw size nine and a half and the two and a half sizes to give room between the toe and the end of the shoe would also give you size twelve for the shoe. You do not need to remember a formula to work any of the problems out. Simply get the basis of the measurements of shoe sizes firmly fixed in your mind. If you have a size stick handy get it and note these points. A size nought is four inches long on the size stick. Each additional inch in length is approximately three whole sizes. Five inches is a size three. Six inches size six, eight inches size twelve and so on. A foot seven inches long measures nine and a half on the size stick and will need nearly an inch or two and a half sizes play between the toe of the foot and the front end of the inside of the shoe. Now you see if you have any one measurement you can get the rest. In the example before you had a foot seven inches long. As size nought begins at four inches the four inches must not be calculated in the measurement for size. They must be subtracted from the seven inches. But as you need practically an inch of play for the toes this must be added, so instead of subtracting four from seven, subtract only three inches which leaves you four inches. As each inch represents three whole sizes multiply four by three and you have the size shoe (twelve) needed to fit a seven inch foot. Simple, isn't it? From seven subtract three and multiply by three.

Now suppose the customer brings in a twig eight inches long which is not the length of the child's foot but the length of the inside of the old shoe. If the old shoe was large enough simply subtract four and multiply by three and you have the size of the shoe needed. In this case it is the same as before or twelve.

In a great many cases customers bring in measurements of the outside of old shoes. It may be surprising to many to learn that one can usually figure the outside measurements of children's shoes the same as the full length of the inside. This is because most all worn shoes have the sole kicked off at the toe, making the outside measurement practically the same as if a twig had been cut the same length as the inside. So the rule of subtracting four and multiplying by three is the one to use.

Brands of Rubbers.

First, Second and Third Qualities—Standard and Differential Brands.

Standard First Quality.	Their Seconds.
Solite.	. . . .
Everstick.	. . . .
Goodyear Glove.	. . . .
Wales Goodyear.	Connecticut.
American.	Para.
Candee.	Federal.
Boston.	Bay State.
Banigan.	Woonasquatucket.
Malden.	Melrose.
Ball Band.	Midland.
Gold Seal.	. . . .
Royal Blue.	Western.
Lambertsville (Stouts).	. . . .
Beacon Falls.	Granite.
Lycoming.	Keystone.
Rex.	. . . .
Educator.	. . . .
Buckskin.	Sunset.

Differential First Qualities.	Their Seconds.
Woonsockets.	Rhode Island.
Victor.	Reliance.
Meyer.	Jersey.
Hood.	Old Colony.
Apsley.	Hudson.
Harvard.	Marboro.
La Cross.	Badger.
Commonwealth.	Security.
Bourn.	Union.
Grand Rapids.	Wolverine.

Third Quality.

Empire.
Colonial.
Middlesex.
Gudvalu.
Central.
Noxall.
Dandy.
Mogul.
Essex.

Cardwriters' Chart

A COMPLETE course in the art of making display and price cards and signs., Beautifully printed in six colors and bronze. Includes specially ruled practice paper. Some of the subjects treated are: First Practice, Punctuation, Composition, Price Cards, Directory Cards, Spacing, Color Combinations, Mixing Colors, Ornamentations, Material Needed, etc. Price, postpaid $1.50.

ORDER FROM

SHOE & LEATHER GAZETTE

1627 Washington Ave.   ST. LOUIS

Successful Advertising—How to Accomplish It

OVER four hundred pages. The cream of the knowledge of men who know how and when to advertise—points retailers should know. Simple ads for every department—the sort to make cash sales over your counter. "Most returns for the least outlay," is the key note. Bound in cloth, size 6¼ x 8½ inches, prepaid, $2.00.

ORDER FROM

SHOE & LEATHER GAZETTE

1627 Washington Ave.   ST. LOUIS

www.ingramcontent.com/pod-product-compliance
Lightning Source LLC
Chambersburg PA
CBHW021148090426
42740CB00008B/995